20 Greatest Astronauts of the World

20 Greatest Astronauts of the World

Monika Koli

Published by
Ocean Books (P) Ltd.
4/19 Asaf Ali Road,
New Delhi-110 002 (INDIA)
e-mail: info@oceanbooks.in

ISBN 978-81-8430-559-3
20 GREATEST ASTRONAUTS OF THE WORLD
by Monika Koli

Edition
2026

Price
₹ 400.00 (Rupees Four Hundred only)

Printed at
Shree Sai Printers, Sahibabad

Contents

1. Where No Man Has Gone Before 7
2. Space Programs: Success and Failures 15
3. Those Who Flew Beyond the Horizon 25
   1. Yuri Gagarin 27
   2. John Glenn 34
   3. Neil Armstrong 41
   4. Sally Ride 49
   5. Buzz Aldrin 56
   6. Chris Hadfield 64
   7. Kalpana Chawla 72
   8. Rakesh Sharma 80
   9. Sunita Williams 87
   10. Alan B. Shepard 95
   11. Christa McAuliffe 102
   12. Jim Lovell 110
   13. Marc Garneau 117
   14. Valeri Polyakov 124
   15. Alexey Leonov 132
   16. Valentina Tereshkova 139
   17. Dennis Anthony Tito 147
   18. Guy Laliberte 153
   19. Guion Bluford 160
   20. Eileen Collins 168
4. Mishaps in Conquest of Outer Space 176
5. The Road Ahead 184

1
Where No Man Has Gone Before

Far off skies speckled with stars have always attracted the mankind. The mysteries hidden in the celestial bodies and beyond them have prodded scientists to explore the unknown. Although, the present generation may not be as inclined towards riding a space shuttle, teenagers and young adults who grew up in the sixties and seventies of the prior century worshipped anyone who entered the space.

The two superpowers of the time, the Union Soviet Socialist Republics (USSR) and the United States of America (USA) and the cut-throat competition of their space programs created an aura around the astronauts. The world treated them as the most fearless heroes who were closest to the fantasy world superheroes such as the Flash, Ironman, Captain America and more.

Star Trek, a famous American TV series, fascinated preteens and teenagers the world over. 'The Starship Enterprise", headed by Captain Kirk and his fleet including Commander Spock and Dr. McCoy, gave many children the dream of following these heroes and becoming astronauts. In fact, NASA made use of the part of the epilogue by Captain Kirk (William Shatner) that came at the beginning of the all star trek episodes. NASA's retrospective on Project Apollo was titled, *Project Apollo: Where No man has gone before: A History of Apollo Lunar Exploration Missions*.

Various Sci-Fi TV series' and smash hits series of space films further glamourized the profession of an astronaut until Howard Wolowitz of, 'The Big Bang Theory', portrayed astronauts as a commonplace human being who may be our next door neighbour.

Today space travels are marketed as pleasure trips; organizations are even offering a chance to the citizens of the earth to buy their own piece of land on other celestial bodies including the Moon and Mars.

The dreams, about the outer space, however, were not as courageous as they are today from the beginning. It took centuries for mankind to actually step into the horizon which it once just watched from faraway with naked eyes. What began as simple observation of stars to understand seasonal cycles or mark directions later developed into a full-fledged science of astronomy; then stepped in astronauts or cosmonauts who left the earth to explore the unknown. These real life superheroes changed the course of the history by daring the human species to discover life outside the horizons.

Astrology: The Earlier Form of Astronomy

People across various civilizations on the earth have always observed the stars planets and other celestial bodies. Most ancient cultures however associated the study of stars and planets or astronomy with astrology. Natural phenomena such as lunar cycle, eclipses, or even rising and setting of the sun carried deep religious meanings. In times when successful crop season was of great importance the astronomer priests held prominent place in the society.

Babylonians, in 1800 BC were able to create a near

perfect calendar, based on their observation of regular movement of the sun and the moon, which tracked the seasonal changes repeated over a certain time period. The Chaldeans or the priest astronomers formed the social class who studied the skies and even made predictions. On the other side of the globe Indian astronomy also developed and the dates for earliest text about the astronomical studies can be traced back to 1400-1200 BCE (with the surviving form of the text, possibly from 700-600 BCE).

Space Travels in Fiction

Slowly astronomy and astrology became entirely two different paths of thoughts. Even though astrology was a prototype of the science form, astronomy developed as the more scientific and logical branch of the two.

Early humans now not only studied the cosmos but also dreamed of touching the stars. Once considered sacred, sun, moon, planets and stars were now viewed as another land mass in outer space where humans could reach. Even before scientists could come up with the schematics of powerful enough spaceships, common man was travelling and discovering the outer space through the writings of the visionary writers who were the actual precursors of the dream of space exploration. Cyrano de Bergerac, the French satirist of the 17th century, gave a fictional account of journeys to the moon and the sun in his works. Histoire comique des états et empires de la lune (1656) and Histoire comique des états et empires du soleil (1662). The book is credited for presenting the first example of a rocket-powered space flight, and for giving scientists the concept of the ramjets, similar to modern day air breathing jet engines. The book relates the account of the narrator's attempt at constructing a way to reach the moon. After a failed attempt the narrator is accidently launched into the space where he meets four legged aliens, musical voices and miraculous weapons that kill and cook the game for a meal at the same time. Here he also meets the ghosts of Socrates, the Greek philosopher and Domigo Gonsales (a character in another science fiction of the time, 'The Man in the Moone'). The narrator also discusses with the ghosts the futility of the concept of God, and how immortality of human beings is an implausible idea as they do not have souls. Next, in his journey to the sun, he uses a machine that operates on solar energy trapped through mirrors. The inhabitants of the sun teach him about the solar system by relating it to how atoms move.

In 1865, French author Jules Verne wrote De la Terre à la Lune (From the Earth to the Moon). Later in the century English novelist and historian H.G. Wells published fictions titled The War of the Worlds (1898) and The First Men in the Moon (1901). Both the writers based these fictions on the sound scientific principles to describe the space travels and encounters with alien beings.

Those Who Made Fiction a Reality

Every action is a result of thought which was once abstract. The twentieth century was the time to give shape to once abstract ideas limited only to the world of fiction. By the middle of the century there were significant developments in the field of science and technology and some countries were able to develop rockets that could send machines, animals and finally humans to the outer space.

Konstantin Tsiolkovsky

Konstantin Tsiolkovsky, the Russian school teacher

and mathematician, was amongst the pioneers to pursue the detailed study of the use of rockets for spaceflight. He published an article, "Exploration of Cosmic Space by Means of Reaction Devices", in 1903 that expounded upon the principles of space flight. During his lifetime Tsiolkovsky published various advanced essays and studies which talked about the theoretical aspects of spaceflight. Although his writings were not backed by practical experiments, his work became the basis of forthcoming researches in the field of space and rocket sciences in the Soviet Union and Europe.

Robert Hutchings Goddard

The space studies in the United States were initiated by Goddard. He became interested in space exploration after reading works such as, The War of the Worlds. In his high school graduation speech (in 1904), he stated that "It is difficult to say what is impossible, for the dream of yesterday is the hope of today and the reality of tomorrow." By 1914 he had already received two patents for rocket technology. In 1919 he published his theoretical essay, A Method of Reaching Extreme Altitudes, in which he claimed that it was possible to launch objects beyond the earth's gravity up to the moon. The innovative idea was not received well and ridiculed widely in The New York Times. However, the publication carried a retraction in its edition of July 17, 1969, the day after the launch of the first manned mission to the Moon. Later he also experimented with liquid-fuelled rockets and his first rocket, launched in Auburn, Massachusetts, on March 16, 1926, rose 12.5 meters (41 feet) and travelled 56 meters (184 feet) from its launching place. He devoted his life towards building and lunching rockets, each rocket excelling in complexity from the previous one. Goddard ideas had limited influence on the development of American rocketry, although early rocket developers in Germany took notice of his work.

Hermann Oberth

Hermann Oberth, a Romanian by birth but a German by nationality was inspired by Jules Verne's, From the Earth to the Moon. The piece of fiction prompted him to study requirements for interplanetary travel. The University of Heidelberg rejected his doctoral dissertation on rocket-powered flight for being too speculative. In 1923 he published Die Rakete zu den Planetenräumen ('The Rocket into Interplanetary Space'), which was based on his thesis. The book included logical explanation of the mathematical theory of rocketry, gave insight into how this theory may be applied to rocket design, and discussed the possibility of constructing space stations and of travelling to other planets. Oberth's work played a crucial role in stimulating the development of rocketry in Germany but he himself had only a limited role in that development. However, Oberth was able to witness his ideas become reality: he was guest at the July 16, 1969, launch of Apollo 11.

Although space exploration was the main motive behind the studies made by the pioneers in this field but the work was put to use for developing military weapons, especially by German military under the rule of Adolf Hitler. The German government created a special development and test center at Peenemünde, located on the coast of the Baltic Sea. Here the team developed various warfare devices including the V-2 (originally called the A-4), ballistic missile. The V-2 was launched successfully in 1942 and was later used on targets in Europe in early September of 1944. The missile was developed as a weapon of war; however the V-2 later served as the predecessor of many of the rockets used in the early space programs of the United States and the Soviet Union.

Space Travels to Serve the Interest of Mankind

The initial years of space exploration were wrought with 'Space Race', between the USSR and the USA with each country trying to surpass the other. In fact, the whole façade of space race was used as the proxy competition for geopolitical rivalries during the Cold War. Even after achieving remarkable success in space endeavours, the two major landmarks of this initial period that still remain unforgettable include the launch of Sputnik 1 on 4 October 1957, the first man-made object to orbit the Earth, and the first Moon landing by the American Apollo 11 Mission on 20 July 1969.

With the advancement of technologies and shift in political ideologies of countries participating in space programs focus moved from one-off flights to renewable hardware, such as America's Space Shuttle Program, and from competition to cooperation with the establishment of International Space Station (ISS).

Although, space travels today are touted as commercial tourism, even in latter half of the twentieth century space odysseys were only commissioned by nations' government due to the very high expenses and security concerns involved. Since, it was the tax payers' money, space explorations were carried out as a series of experiments the results of which would serve the interest of human species in general.

Various Space Programs carried out over the years have been able to achieve what was expected of them in terms of better knowledge of the universe, enhancing a nation's prestige, aiding in military security efforts and other benefits to the human race in general. Private sector profited out from these once unconceivable space adventures. One of the most important of these is using artificial satellites for telecommunications purposes. The beginning of 21st century also heralded the era of commercialized or privately funded space activities; selling space travels as pleasure adventure trip is one of them.

□

2
Space Programs: Success and Failures

Humans are the dominating species on the earth that control and consume a large part of the available natural resources. The intrinsic desire of human beings to explore and regulate the surroundings propelled them to continuously develop the technologies that would help in discovering new extents; outer space was one such area of special interest. The initial study of outer space was carried out by simple but methodical observations of the sky and stars with naked eye. After the invention of telescopes and its future better versions, the sky and its mysteries started unravelling slowly. Physical exploration of space became possible only in the twentieth century when unmanned robotic probes, animals aboard spacecrafts and human spaceflights were sent to far off celestial bodies.

Animals Before Humans in Space

During the experimental stages of space exploration activities unmanned space missions were sent to the space, followed by the spaceships which carried other life forms including fruit flies, fishes, dogs and chimpanzee, wasps, beetles, tortoises, worms, spiders, rabbits, bees, ants, frogs, mice, crickets, rats, newts, snails, urchins, moths, brine shrimp, jellyfish, guinea pigs, butterflies, scorpions and cockroaches. Although, animal activities protested against the investigational use of these naturally trusting, weaker life forms to further the interest of humans in space; scientists called this an important inclusion in the whole procedure. Animals in space gave insight into understanding the impact of microgravity on many biological functions of a living form.

The first life form, that was inhabitant of the earth, to reach space (not including bacteria which are present in earth's atmosphere and may have gone onboard), were the fruit flies. These fruit flies travelled aboard the German V-2 rockets captured by the USA after the Second World War II. The fruit flies served as the subjects to study radiation exposure at higher altitude. Although, the USSR and the USA had been sending to space since 1947, some animal astronauts became as famous as the human astronauts who got to go to space much later.

One such famous animal astronaut was Laika, a mixed breed dog who was put onboard Sputnik 2 on 3rd November, 1957. The dog was rescued from the streets of Moscow and was trained for the space travel. It was argued that since Laika was actually a street dog she was already trained to survive the unknown hardships of lonely travel to space. Laika was taught to live in confined space for longer periods and to eat a nutritious gel for food. Laika's trip to the space was designed to be only one way. It remained unknown for how long Laika could survive in the orbit-few hours, few days or until her life support system gave up. Since the re-entry strategy for the mission was not designed at the time of the launch, Sputnik 2 burned up in the upper atmosphere in April 1958.

After Laika, the Soviet space agency sent two more dogs, Belka and Strelka, into space on August 19, 1960.

Both dogs were actually the first animals to orbit the earth and return alive.

Another celebrity animal astronaut was a chimpanzee named Ham. Ham was launched into space on 31st January, 1961 and was trained to carry out specific tasks during space journey. The name Ham was chosen using the initials of the Holloman Aerospace Medical Center. During the training on earth for its space travel Ham trained to pull levers inside the spaceship which would give him banana pellets. Ham also learned how to avoid electric shocks. He successfully became the first animal to actually interact with a space vessel rather than simply ride in it. Even though landing on the earth was not easy as the capsule drifted away from the expected landing site and Ham had to wait in uncomfortable position for three long hours, he showed no sign of distress upon rescue and even accepted banana and half an orange after his retrieval from the capsule. Ham was sent to the National Zoo in Washington DC. After his brief career as an astronaut, he lived for 17 years and died a natural death in 1983. His grave is in the International Space Hall of Fame in New Mexico.

Early Space Missions

Germany, under the Nazi rule in 1930s and 1940s, set aside special research facilities to explore new age weapons or missiles which were essentially long distance rockets. In World War II, missiles were actually used to attack far away regions. Two hundreds-mile-range V-2 missiles were used to attack the city of London. These missiles travelled at the speed of 3,500 miles per hour reaching the height of 60 miles above the English Channel.

Long distance missiles were the nascent form of future space programs' rockets or spaceships. Following the Second

World War the newly emerged superpowers of the world the USSR and the USA initiated their own missile programs. With the launch of the first artificial satellite, Sputnik 1 on October 4, 1957 by the Soviet Union, the cold war between the two countries took a turn towards space race amongst the two. Not to be left behind, U.S. also launched its first satellite, Explorer 1 into the orbit on January 31, 1958. Next big achievement in space program was again by the Soviet Union when on April 12, 1961, Russian cosmonaut was launched aboard Vostok 1 to orbit the Earth. Vostok 1 was able to reach an altitude of 327 kilometers and the flight lasted for 108 minutes. The flight was soon followed by the USA astronaut Alan Shepard on May 5, 1961, when he piloted the Mercury-Redstone 3 Mission to space and became the second person, and the first American, to travel into space. However, the flight in many ways was not at par with the orbit journey of the Russian cosmonaut Yuri Gagarin. Alan's Mercury Spacecraft 7, Freedom was able to achieve the height of only 187 km and remained in the sub orbit for only 15 minutes. Finally on 20 February, American cosmonaut John Glenn became the first American to orbit the Earth.

The USA by now was looking for 'the first', in the space race. President John F. Kennedy in 1961 announced a national goal of, "Landing a man on the moon and returning him safely to Earth within a decade." NASA, the US space agency fulfilled the nation's dream and the astronaut Neil Armstrong successfully landed on the moon on July 20, 1969. After this, several space expeditions were carried out, some of which were successful and some not so successful, while others ended up in tragic disasters, such as the explosion of the spaceshuttle Challanger, 73 seconds into the flight in 1986 and later disintegration of spaceshuttle

Columbian in 2003. All crew members were killed in both the accidents, forcing America to work towards putting in place uncompromised safety procedures in place.

In early 1970s navigation satellites were used for civilian purposes and mariner spacecraft was already in the orbit of Mars. By the end of the decade, scientists on earth had received detailed pictures of Saturn and Jupiter as well as of their rings and moons (natural satellites of the planets), through the spacecraft the Voyager. The Apollo Soyuz Test Project brought together American and Russian space technologists and the mission became the first internationally crewed space mission.

In the 1980s it became possible to transmit television signals through satellites. Man-made satellites also discovered a hole in ozone layer over Antarctica. The images sent back to the earth by satellites could be used for indicating forest fires before they were uncontrollable. Through satellites only it was possible to procure the pictures of Chernobyl nuclear powerplant disaster in 1986.

Images from artificial satellites offer great aid to warning countries and could be used to save many innocent lives. The Gulf War proved this further when conflicting countries used their space resources to gain information of the enemy country's troop location and formation, probable missile attacks, and to help navigate across the otherwise featureless topography of the desert land. The war was put to end quickly and gave decisive advantage to one of the sides; saving many lives.

Modern day space missions are not only limited to the single cause of sending human into the space for the sake of sending them there but the information collected from the high grounds can be put to use for various purposes.

With the establishment of the International Space Station the low Earth orbit, the message is clear that space exploration activities will benefit the human race in general only if these are carried out in cooperation with each other.

Early Years Achievements in Space Programs Which Paved the Road for Future Space Missions

Mission details	Date accomp-lished on	Name of the spacecraft/ satellite/ launch vehicle	Country/ agency involved
First artificial satellite of the earth was launched	4 October, 1957	Sputnik 1	Union of Soviet Socialist Republics
First animal (dog Laika) to orbit the earth	3 November, 1957	Laika went into space aboard Sputnik 2	Union of Soviet Socialist Republics
First spacecraft to hard-land on the Moon before manned spaceflight reached its surface	14 September 1959	Luna 2	Union of Soviet Socialist Republics
The spacecraft sent home the very first pictures of the far side of the Moon	7 October, 1959	Luna 3	Union of Soviet Socialist Republics
The first of its kind application satellite was launched for weather observation	1 April,	TIROS 1 (Television Infrared Observational Satellite)	United States of America

First successful recovery of a payload from the earth's orbit	11 August, 1960	Discoverer 13 (part of Corona reconnaissance satellite program)	United States of America
Soviet cosmonaut Yuri Gagarin was the first person to orbit the earth	12 April, 1961	Vostok 1	Union of Soviet Socialist Republics
Scientists were able to receive the first data from another solar planet, Venus	14 December, 1962	Mariner 2	United States of America
Russian woman cosmonaut Valentina Vladimirovna Tereshkova became the first woman to enter space	16 June, 1963	Vostok 6	Union of Soviet Socialist Republics
First telecom-munications satellite was launched to operate in geostationary orbit	26 July, 1963	Syncom 2 (synchronous communication satellite)	United States of America
First space walk outside the space -craft was accomplished by Aleksey Leonov	18 March, 1965	Voskhod 2	Union of Soviet Socialist Republics
First pictures of the planet Mars were sent back to the earth	14 July, 1965	Mariner 4	United States of America

First spacecraft to soft-land on the Moon or any other planetary body apart from the earth	3 February, 1966	Luna 9	Union of Soviet Socialist Republics
Cosmonaut Vladimir Komarov died due to parachute failure of the descent module while landing. This was the first in-flight fatality in the history of space flight	24 April, 1967	On Soyuz 1	Union of Soviet Socialist Republics
Frank Borman, James Lovell, and William Anders became first humans to orbit the Moon	24 December, 1968	Apollo 8	United States of America
Neil Armstrong became the first human to walk on the Moon	20 July, 1969	Apollo 11	United States of America
By an unmanned spacecraft returned First lunar samples successfully	24 September, 1970	Luna 6	Union of Soviet Socialist Republics
Venera 7 travelled into the space to make the first soft landing on another planet (Venus)	15 December, 1970	Venera 7	Union of Soviet Socialist Republics

Launch of the first space -station	19 April, 1971	Salyut 1	Union of Soviet Socialist Republics
NASA's unmanned space probe became the first spacecraft to orbit a solar planet (Mars) other than the earth	13 November, 1971	Mariner 9	United States of America
Mars 3 became the first spacecraft to soft-land on Mars	2 December, 1971	Mars 3	Union of Soviet Socialist Republics
American spacecraft was the first to fly by Jupiter	3 December, 1973	Pioneer 10	United States of America
First successful international docking in space happened during Apollo-Soyuz Test Project	17 July, 1975	Apollo and Soyuz	USA and U.S.S.R.
First reusable spacecraft launched and returned from space	April 12-14, 1981	Space shuttle Columbia	United States of America

Source: www.britannica.com

□

3

Those Who Flew Beyond the Horizon

YURI GAGARIN
(The First Man in Space)

In the year 1961, human civilization stood proud. A mission impossible was achieved; Yuri Alexeyevich Gagarin became the first human to escape the gravity and orbit the Earth in spaceship Vostok 1. Gagarin came back a hero after spending 108 minutes in space orbiting the earth at 18,000 miles per hour at an altitude of 302 kilometers.

Gagarin's only tryst with space made him an international hero. Till date he remains the most decorated astronaut with medals and honours from at least twenty-five countries. Gagarin died in a tragic air crash of MIG-15UTI, which he was piloting during a routine training flight, on 27 March 1968.

Yuri Gagarin as a Child and Teenager

Gagarin was the third born of his parents Alexey Ivanovich and Anna Tiofeyevna Gagarina. He was born on 9 March, 1934 in Klushino, a small village 100 miles in west of Moscow. Gagarin's father was a carpenter, cabinetmaker, farmer and bricklayer while his mother was milkmaid. His parents worked on a KolKhoz or a collective farm. According to Soviet social standards Gagarin's heritage was flawless. During World War II Gagarin's family also suffered Nazi occupation like millions of the people across Soviet

Union. When Gagarin was only seven, in November 1941, his village was occupied by Nazis during an army advance on Moscow. Gagarin family was made to leave the house as an officer took over their residence. Gagarin and his family lived in a 10 feet by 10 feet mud hut, which they were allowed to build on the land behind their house. The family spent a little less than two years in this dugout until the occupation ended. His older siblings were made to deport to Poland for slave labours in 1943 by Nazi army and did not return until after the war was over in 1945. Later Gagarin shared in his recollections that how he and his friends made trouble for Nazis.

In 1946, Alexey Ivanovich and Anna Tiofeyevna took their family to Gzhatsk where Gagarin continued his secondary education. The town of Gzhatsk was renamed Gagarin after his death in 1968. During the six grades of secondary school, Gagarin studied his favourite subject mathematics and physics. It was during these formative years, Gagarin became acquainted with the idea of flying planes and achieving the extraordinary. A teenaged Yuri and his friends once saw a Russian Yak fighter plane. Gagarin make a force landing in an open field near his home while returning from battle. Its wings were damaged because of the bullets. The medal covered pilots emerging from the plane created an everlasting impression on Yuri's young mind. Later he was recorded saying,

"We understood immediately the price that had to be paid for military decorations. We boys wanted to be brave and handsome pilots. We experienced strange feelings such as we had never known before."

Career and Marriage

In 1950, a 16 year old Gagarin joined as an apprentice foundryman at a steel plant near Moscow and also enrolled for seventh grade classes in the evening at the local, "young workers", school. During these years he read the works of Konstantin Tsiolkovsky, the Russian rocket pioneer. Completion of seventh grade along with a year and half at the trade school (where trained to become a foundryman), was closely followed by his selection for a four year training course at Industrial Technical School in Saratov.

In the fourth year of his school, his teenage dream of becoming a pilot gained wings when he was offered a chance of joining a local flying club and train as a Soviet air cadet. Gagarin was a hardworking and intelligent student. Not only he worked hard to earn extra money by working part time as a dock labour on the river Volga, he soon took his solo flight in 1955. Gagarin was often praised for his skills at handling planes and smooth landing. Gagarin's mentor and flying instructor Dmitry Pavlovich Martyanov impressed by his quick leading soon predicted, "He'll make a wonderful pilot."

Gagarin also aced the art of parachuting while learning to fly, a skill that came in handy when he had to eject from the Vostok 1, as it re-entered the earth's atmosphere. So enthusiastic he was about flying that he willingly spent a summer flying and living in a tent pitched near an airfield situated outside Moscow. His solo flight came in the year 1955 when he flew Yak-18 trainer.

"That first flight filled me with pride and gave meaning to my whole life," said later.

After graduating from the technical school, Gagarin joined Soviet Air Force following the advice and recommendation of Martyanov. He was sent to Orenburg to attend First Chkalov Air Force Pilot's School and by 1957 he was taking solo flight in MiG-15. During his time in Orenburg, he met his future wife Valentina Ivanovna Goryacheva or Valya.

Gagarin, now 23, graduated with top honours from the Orenburg Aviation School on 7 November 1957, becoming a Lieutenant in the Soviet Air Force. The day of his graduation was also the day he married shy and beautiful Valya.

As a fighter pilot he was first posted at Soviet Union's airbase in the arctic where the weather conditions were always unfavourable for flying. The Luostari airbase in Murmansk Oblast was close to the Norwegian border and Gagarin made his flights in the light of the aurora borealis (northern lights). By 1959 Gagarin had already reached the rank of Senior Lieutenant.

Training as a Cosmonaut

Soviet's Luna-3, an automatic interplanetary station, was launched in 1959. The spacecraft transmitted images of the lunar far side for the first time. Sputnik's historical flight to space was a significant news and so was Nikita Khrushchev's announcement of Soviet's intentions to soon send a manned spaceflight into outer space.

The adventure of space program always attracted Gagarin. He applied for cosmonaut training and after a rigorous selection process was along with 19 other pilots for Soviet's next chapter in space program. All participants were moved to Star Town outside Moscow which was built especially for the cosmonauts.

The training process was demanding where he and other candidates had to undergo arduous experiments designed to test their physical and psychological endurance. The candidates were kept in sensory deprivation chambers for long periods, tested for heat endurance, experienced weightlessness and took test flights under stress amongst other experiments and tests that were run on them. Their reactions were monitored. Through all this Yuri always remained calm and resolute.

His fellow cosmonauts training participants described him as a, "fit comrade, never loses heart, a man of principles, bold and steadfast, modest and simple, decisive, a leader."

These qualities made Gagarin an obvious choice for next level training program, where only six of the earlier participants made their way. This elite training group called Sochi Six. Out of Sochi Six, the first cosmonaut of Vostok program was to be selected. Finally, Gagarin and Gherman Titov were chosen based on their performance during experiments and training sessions as well as their physical characteristics. Both were rather short. Vostok cockpit was small and could fit only a short statured human. Gagarin at 5 ft 2 in became the ultimate choice.

Vostok 1 Mission

Gagarin, onboard Vostok 1 spacecraft, took off from Baikonur Cosmodrome. Vostok I remained in space for 108 minutes, reached the maximum height of 327 kilometers and orbited the Earth once. Somewhere over Africa, Gagarin's travel back to the Earth began. The spacecraft carried extra provisions which could last for 10 days, if the engines failed and Gagarin was forced to wait till the natural decay of the orbit. However, this was rendered unnecessary and Gagarin re-entered Earth's atmosphere. The forces he experienced at this time were eight times the gravitational pull of the

Earth. The Design of Vostok 1 did not include engines that could slow down the spaceship as it would re-enter the Earth. A little secret that was not revealed until 1971 by Soviet leaders was that Gagarin had actually ejected himself at the height of seven kilometers above the earth and parachuted down. The governing body for aerospace records, Fédération Aéronautique Internationale (FAI), had laid down the condition that a mission would be counted as an official spaceflight only if the pilot lands with the spacecraft. Soviet leaders hid this fact. Despite this minor contortion of facts, Gagarin still paved the way for mankind into outer space.

Gagarin recorded his experience in space, in his post-flight report as:

> "The feeling of weightlessness was somewhat unfamiliar compared with Earth conditions. Here, you feel as if you were hanging in a horizontal position in straps. You feel as if you are suspended."

Life after Vostok 1

Gagarin's successful space journey marked Soviet Union's victory in the space program race with the USA. Newspapers world-over carried his photographs, bio-graphy and details of his space journey. Gagarin received numerous medals and awards from various nations. He was conferred the title of the Hero of Soviet Union, the highest award of the Soviet Union, in a pompous ceremony held in Kremlin.

Gagarin travelled widely visiting many countries including Brazil, Canada, Egypt, Finland, Germany, Japan and UK to promote Soviet union's space program and its successful missions. He was welcomed and received with warmth everywhere despite of the cold war between the two superpowers of those days.

Gagarin was not to sit on his laurels for too long and by 1962 returned to his base in Star City where he worked on designs for making reusable spacecrafts. The Soviet government tried to keep him off the flights, apprehensive of losing their national hero in any mishap.

Tragic End

In the December of 1963, Gagarin became deputy training director of the Star City cosmonaut training base. Two years later he was re-elected to continue with his position. He could not keep himself away from his love of flying for too long and proceeded to re-qualify for being a fighter pilot. He met a tragic end at the young age of 34, on 27 March, 1968 when the MiG-15UTI, he and flight instructor Vladimir Seryogin were flight testing, crashed. Gagarin and Seryogin were cremated and their ashes buried in the Kremlin Wall on Red Square.

In Gagarin's death the entire world lost a star. Gagarin was an international hero who inspired many youngsters to become cosmonauts. His smile was his one of his attractive traits apart from his affable personality as a public hero.

Sergei Korolev, one of the masterminds behind the early years of the Soviet space program, later said that Gagarin possessed a smile "That lit up the Cold War."

Gagarin, who lived the life which others only dreamed of, was equally decorated in his death. Every year, 12 April, Gagarin's night is marked for international celebrations to commemorate and honour the mile-stones in space explorations.

□

JOHN GLENN
(The Oldest Person to Fly to Space)

John Herschel Glenn, was born to John Herschel and Clara Sproat Glenn on July 18, 1921 in Cambridge, Ohio. He became the first American to orbit the Earth and his second journey to space made him the earliest born American to go to space and overall the second earliest born person to be in space, next to Soviet cosmonaut Georgy Beregovoy. In 1978 Glenn was honoured with the prestigious Congressional Space Medal of Honour. In 1990 he was inducted into the U.S. Astronaut Hall of Fame. He is the last surviving member of the, "Mercury Seven" The group, "Mercury Seven", included military test pilots selected by NASA in 1959 to train as America's first astronauts and fly the Project Mercury Spacecraft. On February 20, 1962, Glenn was launched into space aboard Mercury's Friendship Seven Capsule. He orbited the Earth three times in this journey at a speed of more than 17,000 mph, which lasted almost five hours. America's previous spaceship carrying astronauts Shepard and Grissom made only sub-orbital journey, i.e. they had not travelled all the way around the Earth, this made John Glenn America's first man to orbit the Earth.

John Glenn's parents moved to New Concord when he twas two. John's father ran a plumbing business. The Glenns also built a home in New Concord, where students from nearby Muskingum College Boarded. Glenn's early years were spent in the company of older students. He was tutored by his devoted mother and always encouraged by his father who liked to travel.

"A boy could not have had a more idyllic early childhood than I did," Glenn reminisced later in his writings.

John studied in New Concord High School and after that in 1939 went to Muskingum College to study Engineering. In 1941 he garnered a private pilot's license to get credit in a physics course. John could not complete his senior year in residence, nor did he take the proficiency exams; both of which were required by the school to award Bachelor of Science degree. Later in 1962 however, the school did grant John Glenn his degree, after his first space mission.

During the time when United States embroiled into World War II after the Pearl Harbor attack by Japanese, Glenn quit college and enlisted in the U.S. Army Air Corps. However, he never had a chance to be called for duty. In March 1942, Glenn enlisted for Naval Aviation Cadet Program. John was sent to the University of Iowa, where he too took preflight training, followed by his primary training at NAS Olathe, Kansas. Later at the NAS Corpus Christi, during advanced training, he opted for transfer to the U.S. Marine Corps. Glenn's training was completed in 1943 and thereafter his duty was to fly R4D transport planes as a part of the Marine Squadron VMJ-353. Later he was assigned to VMF-155 and flew F4U Corsair fighter pilot. His career as an aircraft pilot was full of adventures. He flew 59 combat missions in the South Pacific during World War II. Again he asked for combat duty during Korean conflict and flew 63 missions during the time. During the two wars he flew a total of 149 missions and was awarded various medals including the Distinguished Flying Cross which he was awarded six times. Glenn has nearly 9,000 hours of flying time, with approximately 3,000 hours in jet aircraft.

Glenn afterwards returned to graduate from the U.S. Naval Test Pilot School. After his graduation in 1954, he was inducted as an armament officer. His work included flying planes to high altitude and test their cannons and machine guns. He became a part of the Bureau of Naval Weapons of Aeronautics (then the Fighter Design Branch of the Navy Bureau). During this tenure (November 1956 to April 1959), he worked as a test pilot on Navy and Marine Corps Jet Fighters. He also attended the University of Maryland during the same time.

The daring pilot flew the first supersonic transcontinental flight in the jet aircraft Vought F8U-3P Crusader. He went from coast-to-coast in 3 hours, 23 minutes and 8.3 seconds, starting from NAS Los Alamitos, California and landing in Floyd Bennett Field, New York.

An interesting story linked with Glenn's courageous flight relates that as Glenn flew over his hometown, a child in the neighbourhood ran around shouting, "Johnny dropped a bomb! Johnny dropped a bomb! Johnny dropped a bomb!" The sonic boom had actually shaken the town. The mission was called the Project Bullet. The mission included the first transcontinental flight averaging with supersonic speed and the first continuous transcontinental panoramic photograph of the United States.

The United States was locked in a heated space race with the Soviet Union in the 1950s. In 1958, NASA the newly formed body to further the space program of the U.S.A., began recruiting astronauts. The conditions included that the all space aspirants had to be military test pilots with specified flight hours, aged between 25 and 40, not above the height of 5 feet 11 inch and should hold a degree in a scientific field. Out of the 508 aspirants, who were all put to various mental and physical tests, the final 'Mercury Seven', included John Glenn and six others (Alan Shepard, Gus Grissom, Scott Carpenter, Wally Schirra, Gordon Cooper, and Deke Slayton). NASA introduced these seven astronauts to public in April 1959 in a press conference. Out of the seven astronauts Glenn just barely came near the requirements, he was nearing 40 and had not completed his science degree. John served as the backup pilot for the US previous space mission which took Alan Shepard and Gus Grissom for their sub orbital journey in space. On February 20, 1962, Glenn was launched into space aboard Friendship Seven as a part of Mercury-Atlas six mission. He circled the Earth three times in the timeframe of 4 hours, 55 minutes, and 23 seconds.

"Zero G, and I feel fine," were Glenn's first words on becoming weightless upon reaching orbit.

Citizens of Perth, Western Australia showed their enthusiasm when Glenn passed overhead and turned on the lights in their houses, cars and street, making the city what was known as the, "City of Light", world-over. The city recreated the same phenomenon when Glenn went to space again in 1998.

Glenn had always been a courageous test pilot, achieving daring feats while in service; his first mission to space was also not without a little misadventure. The flight controllers at NASA headquarters suspected that the heat shield of Glenn's capsule had loosened and will not work properly as Glenn would re-enter the atmosphere causing the capsule to burn up. Glenn was instructed to keep the retrorocket pack above the shield so the shield could be kept in place. He made these readjustments and landed safely. However, later it was determined that it was false alarm and the indicator was faulty.

Glenn's service to the US in advancing its space program did not go unnoticed and he was greeted with numerous accolades in coming years. He was greeted by the then President Kennedy three days after his flight to space and had the honour of receiving New York ticker-tape parade, a special parade organized by city officials on triumphal occasions. Glenn's ticker-tape parade was one the longest and largest parade, the only other parade of this stature was held in the honour of World War II and Korean War General Douglas MacArthur in 1951.

While training with NASA, the psychologists there had observed that he was the astronaut with attributes best suited for public life. Glenn and Kennedy family were already friends, after their first meeting when President Kennedy awarded John Glenn with the NASA Distinguished Service Medal. Attorney General Robert F. Kennedy suggested Glenn that he should contest in Democratic primary election to be held in 1964. Glenn too liked the idea and announced his resignation from the space program to run the coming elections. However, he had to withdraw from the elections owing to an accident, in which he hit his head on a bathtub and developed a concussion and injured his inner ear. Glenn also had to apply for convalescent leave from the Marine Corps, till the time he could fully recover. In 1965, he retired from the Corps as a Colonel and became a business executive for Royal Crown Cola. Glenn and the Kennedies remained close and was accompanying Robert Kennedy on the day he was assassinated in 1968. Glenn lost another chance to hold an elected political post in 1970, this time due to losing the Democratic primary election against fellow Democrat Howard Metzenbaum. Again in 1974, Glenn challenged Metzenbaum and after winning the primary election, went on to win the general election too. This began John Glenn's career as a Senate that continued till 1999.

It is said that what won Glenn his primary elections was the speech that came to be known as the, 'Gold Star Mothers', speech. The speech challenged his opponent's statement that Glenn's military and astronaut credentials were not worthy enough as he, "Never held a payroll." Glenn, in his memorable reply said that his opponent, Metzenbaum should:

> "Look those men with mangled bodies in the eyes and tell them they didn't hold a job. You go with me to any Gold Star Mother and you look her in the eye and tell her that her son did not hold a job."

The Ohio Democrat, Glenn served four terms in Congress. He held various posts over these years on several committees. He was main author of the Nuclear Non-Proliferation Act of 1978, chaired the Committee on Government Affairs from 1987 until 1995 and was a member of Foreign Relations and Armed Services committees and the Special Committee on Aging. As a politician Glenn was considered be quite outspoken on many issues. In 1984, he made a bid for the Democratic presidential nomination but was not successful.

Glenn also campaigned for more funds for space exploration, science and education. Glenn was depicted as a "zealous moralizer", in the book "The Right Stuff", by Tom Wolfe. The book, about the "Mercury Seven" astronauts, was a best-seller and was later made into a feature film of the same name.

Glenn's second chance to fly to space was on October 29, 1998. He was still a sitting senator at that time and aged 77. He flew aboard Discovery Mission STS-95 as a Payload Specialist. Glenn later stated in his memoirs that he was ignorant of the fact that NASA was even considering his name for another chance in space until the decision was made public. Glenn's second flight to space was considered especially useful as it offered a valuable insight upon the effect of weightlessness and other aspects of space expeditions on the same person at two points in life-at an interval of 36 years in-between. The inflight studies also gave information about the effects of spaceflight and weightlessness on the elderly.

The crew of STS-95 was honoured with ticker-tape parade after their successful nine day mission in space. This made Glenn tenth, and latest person to receive multiple ticker-tape parades in their lifetime. An active public figure, Glenn strongly opposed Dennis Tito's space flight to the International Space Station for the reason that this visit was not of any use from the point of view of science.

John Glenn donned multiple hats over the years, that of Marine, pilot, astronaut, corporate executive, and a senator. In all roles his passion and zeal made him outstanding. After retirement from the senate and success of his second journey to space, John and his wife Annie Glenn founded the John Glenn Institute for Public Service at the Ohio State University. The Institute offers programs aiming at the improvement of public service quality and inspire more and more young people to pursue careers in government. The Glenns also continue to serve as trustees of Muskingum College, their alma mater.

□

NEIL ARMSTRONG
(The First Man to Step on Moon)

Neil Alden Armstrong was born in Auglaize County, Wapakoneta of Ohio state, USA. His father Stephen Koenig Armstrong was an auditor and worked for the Ohio state government. Stephen and his wife Viola Louise Engel welcomed their first born in their home on August 5, 1930 and named him Neil Alden Armstrong. In his growing years Neil lived in at least 20 places as his father's work took him around the state frequently. Neil Armstrong from the age of two, he accompanied his father to the Cleveland Air Races. It was during these impressionable years that his love for flying grew. His first airplane flight was on 20 July in 1936, when at the age of 5, Neil along with his father rode in Ford Trimotor, popularly called 'Tin Goose'.

In the year 1944 the Armstrong family moved back to Neil's birthplace Wapakoneta. Here he went to Blume High School and started with his flying lessons. Neil was only 16 when he achieved the student flight certification even before he could get his driver's license. He also made a solo flight soon after his birthday in the month of August the same year. All through his schooling years Neil actively participated in Boy Scouts. He also reached the rank of Eagle Scout. Later as an adult the Boy Scouts of America decorated him with Distinguished Eagle Scout Award and Silver Buffalo Award.

Armstrong too was proud of his association with the Boy Scout Society and sent them greetings while onboard Columbia heading towards Moon, on July 18, 1969. He said:

> "I'd like to say Hello to all my fellow Scouts and Scouters at Farragut State Park in Idaho having a National Jamboree there this week; and Apollo 11 would like to send them best wishes."

World Scout Badge was part of Armstrong's few belongings which he carried in his journey to the Moon. Naturally interested in aviation, Armstrong decided to pursue a degree in Aeronautical Engineering from Purdue University. His college tuition was paid for under the Holloway Plan. Successful applicants committed to two years of study, followed by two years of flight training and one year of service in the U.S. Navy as an aviator, then completion of the final two years of their bachelor's degree. Candidates had to promise to not marry until graduation, signed the "Aviation Guarantee", to serve on Active Duty for at least four years, and would not receive a promotion to Ensign until two years after they received their Midshipman's warrant.

However, could not finish his two years of college and at the young age of 18, was called to join Navy on January 26, 1949. Neil underwent flight training at Naval Air Station Pensacola for 18 months and qualified as a carrier lander aboard US navy's aircraft carriers USS Wright and USS Cabot.

On 16 August, 1950 shortly after his 20th birthday Neil Armstrong became a fully qualified Naval Aviator. He made his first flight in an F9F-2B Panther, a jet in the US

Navy fleet on January 5, 1951. The Holloway Plan under which Armstrong's education was funded stipulated that the recipient of the scholarship would not be promoted to Ensign until two years after receiving Midshipman's warrant. However, within the week after he made his first jet carrier landing on USS Essex he was promoted from Midshipman (the juniormost rank in the US Navy), to Ensign.

During the Korean War, Armstrong was a part of at least 78 combat missions as research pilot. Battle ridden skies of Korea put Armstrong through various dangerous situations, which he faced with sheer courage and presence of mind. In one of the war incidents, Armstrong was piloting F9F Panther and while trying to make a low bombing run his plane was hit by anti-aircraft fire. In his attempt to regain control the plane rammed into a pole resulting in severe damage to the right wing of Panther. Armstrong still was able to manoeuver the plane to friendly territory, but ejection was his only option as safe landing was not possible with damaged wing. He tried to eject over water where he could be later rescued by Navy helicopters but the impact made his ejection seat to blow back over land. His roommate from flight school picked and drove him to the base. The wreckage of his plane remained untraced until later.

Armstrong was decorated with various medals for his exemplary valour during Korean War. He flew 78 missions spread 121 air hours during this war in the January of 1952. He was awarded air medals for 20 of these combat missions, for next 20 missions he received a Gold Star along with the Korean Service Medal and Engagement Star. At the age of 22, Armstrong left active Navy life and joined U.S. Naval Reserve, as a Lieutenant, Junior Grade. He served the Naval Reserve for eight years and resigned from his commission on October 21, 1960.

Armstrong returned to Purdue to complete his degree. This time he met Janet Elizabeth Shearon, a student of home economics. The couple got married on January 28, 1956, at the Congregational Church in Wilmette, Illinois. They had three children, Eric, Karen and Mark. Their daughter Karen was diagnosed with a malignant tumour of brain. Her health deteriorated fast and she died at the age of two. While Janet could never finish her degree, Armstrong went on to attend University of Southern California and completed Master of Science in Aerospace Engineering in the year 1970. Later he was felicitated with honorary doctorates by several universities.

After his graduation from Purdue, Armstrong applied at the National Advisory Committee for Aeronautics (NACA) High-Speed Flight Station at Edwards Air Force Base for the position of test pilot. Committee had no open positions at that time but chose to forward the application to the Lewis Flight Propulsion Laboratory, Cleveland. Armstrong joined the lab in March 1955 but within a couple of months found a new job at Edwards Air Force Base.

Armstrong's adventures with flying continued here and he was involved in several incidents which became part of Edwards folklore. These incidents were corroborated and find mention in various chronicles and memoir of his colleagues.

In one such incident on March 22, 1956, Armstrong and Stan Butchart were flying a Boeing B-29 Superfortress. The plane had to airdrop a Douglas D-558-2 Skyrocket. While flying Boeing B-29 Superfortress, one of it four engines malfunctioned. With the faulty engine the pilots could not have reached the required height they needed to launch its Skyrocket payload and also could not land safely with the skyrocket still attached to the planes belly. Armstrong and Butchart managed to increase the speed of the aircraft by bringing the aircraft into a nose-down alignment and launched the payload. But by now the faulty engine had disintegrated and in the process damaged the other engines. Butchart and Armstrong were able to land safely making a slow, circling descent from the height of 30,000 ft using only one engine.

In another incident on April 20, 1962, Armstrong was assigned to test a self-adjusting control system of X-15. This was his sixth flight on the plane, which could attain very high speed. Armstrong reached the height of 207,000 feet (63 km) but during descent bounced off the atmosphere back up to 140,000 feet because the aircraft's nose had been held up for too long while coming down. At the high altitude, to which the plane was thrown back, aerodynamic surfaces were of no use because of the air being very thin. However, with the ability to come out of difficult situations. He was able to slowly descend flying the plane at the height of 100,000 feet (30 km) and safely bring it back to the landing area. This was recorded as the longest flight X-15 ever made in terms of both time and distance from the ground track.

Armstrong flew many high speed aircrafts including X-15 rocket plane which could travel at the speed of 4000-mph. During his association with Edwards, which later became part of NASA, he flew more than 200 different models of airplanes, which included rockets, jets, helicopters and gliders.

Armstrong was selected in 1958 for, Man In Space Soonest Program of the U.S. Air Force. He served as a part of the pilot consultant group for an underdevelopment military space plane. Later he and six other pilot-engineers were selected by the U.S. Air Force to fly the space plane when it was ready.

NASA also sought applications for its second group of astronauts. Armstrong also applied for the Apollo Program as a daring pilot at heart, he always wanted to investigate new aeronautical environments. On September 13, 1962, he was asked to join the NASA Astronaut Corps. Even though, the names of those selected were kept a secret, newspapers were still abuzz with the news of Armstrong as the 'First Civilian Astronaut'.

Armstrong was assigned as command pilot for the Gemini 8 mission, which was launched on March 16, 1966. Armstrong performed the first successful docking of two vehicles in space. However, all of the mission's objectives were not achieved and Armstrong along with Davis Scott, the command pilot of the mission, had to cut short their journey and return to earth. Following this Armstrong was also a part of Gemini 11 mission which was considered successful, where Astronauts Charles 'Pete' Conrad, Jr. and Richard F. Gordon, Jr. performed their assigned tasks in space while Armstrong served as CAPCOM (capsule communicator) from NASA, Houston. Following this Neil also served as the backup commander for Apollo 8 mission and was offer the commandership of NASA's next historical mission Apollo 11.

Finally, with the Apollo 11 mission, Armstrong's got his chance to create history. American Astronauts Armstrong and Buzz Aldrin landed on moon on July 20, 1969.

Apollo 11 was launched by Saturn V rocket from Kennedy Space Center in Merritt Island, Florida, on July 16 and was the fifth manned mission of NASA's Apollo program. The Apollo spacecraft had three parts: a Command Module (CM) with a cabin for the three astronauts, and the only part that landed back on Earth; a Service Module (SM), which supported the Command Module with propulsion, electrical power, oxygen, and water; and a Lunar Module (LM) for landing on the Moon. Armstrong announced the landing to Mission Control and the world with the words, "Houston, Tranquillity Base here. The Eagle has landed."

Armstrong became the first person ever to step on the moon. With his first step he also spoke the famous words: "That's one small step for [a] man, one giant leap for mankind." Armstrong spent about two and a half hours outside the spacecraft, and together with Aldrin collected 21.5 kg of lunar material for return to Earth. The third member of the mission, Michael Collins, piloted the command spacecraft alone in lunar orbit until Armstrong and Aldrin returned to it just under a day later for the trip back to Earth.

Together Armstrong and Aldrin performed various activities on moon, which were planned beforehand. They unveiled a plaque commemorating their flight, and planted the flag of the United States on the moon's surface. Shortly after, President Richard Nixon spoke to the Astronauts. Majority of photographic tasks were performed by Armstrong with a single Hasselblad camera. Therefore, there are only five images of Armstrong partly shown or reflected. In 2013, Popular Science's Photo Gallery included a photo that Armstrong took of Aldrin but his own image is visible on Aldrin's helmet as one of the best astronaut selfies.

The Apollo 11 craft landed back on July 24, 1969 and came down in the Pacific Ocean west of Hawaii. The crew and the craft member were picked up by the U.S.S. Hornet, and the three astronauts were put into quarantine for three weeks. Apollo 11 astronauts were given a warm welcome home. New York city streets were lined by the crowd, which their heroes as they were honored in a ticker-tape parade. Later Armstrong received many awards, including the Medal of Freedom and the Congressional Space Medal of Honour. Despite being one of the most famous astronauts in history, Armstrong largely shied away from the public eye. Even in his final years, Armstrong remained committed to space exploration. The press-shy astronaut returned to the spotlight in 2010 to express his concerns over President Barack Obama's decision to cancel the Constellation program, which included another mission to the moon.

"America is respected for its contributions it has made in learning to sail on this new ocean. If the leadership we have acquired through our investment is simply allowed to fade away, other nations will surely step in, where we have faltered. I do not believe that would be in our best interests."

Armstrong was admitted to the hospital for a heart bypass operation in August 2012. However, due to complications which developed post-surgery Armstrong died on August 25, 2012, at the age of 82 in Cincinnati, Ohio.

> A statement released by his family after his death read: "For those who may ask what they can do to honour Neil, we have a simple request. Honour his example of service, accomplishment and modesty, and the next time you walk outside on a clear night and see the moon smiling down at you, think of Neil Armstrong and give him a wink."

□

SALLY RIDE

(The First American and the Youngest Woman to Travel to Space)

Sally Kristen Ride holds the record for being America's first women that flew to space. She was a physicist and an astronaut. Sally Ride was born in Los Angeles. She joined NASA in 1978 and in 1983, at the age of 32, went to space aboard the spaceshuttle Challenger. She was also the youngest American astronaut to travel to space. She made two journeys to space on the Orbiter Challenger and later left NASA in 1987 to pursue her career in academia. She worked at her *alma mater* Stanford University, at their center for International Security and Arms Control. Later she joined as a professor of physics at the University of California, San Diego. Here her time was primarily spent in researching non-linear optics and Thomson scattering. Sally, however remained associated with NASA in some way or the other while serving on two different committees that investigated the Challenger and Columbia space shuttle disasters.

Sally Ride was born on May 26, 1951, in California to Dale Burdell Ride and Carol Joyce. Sally's father was a political science professor at Santa Monica College, while her mother worked as a volunteer counsellor at a women's correctional facility. Both parents were elders in the

Presbyterian Church. Her elder sister Karen 'Bear' Ride is a Presbyterian minister. Sally Ride went to Portola Junior High and Birmingham High School in her initial years of schooling. Later under a scholarship program, she attended and graduated from the private Westlake School for Girls in Los Angeles. Sally was always interested in science and also played tennis. Her competitive streak was evident on court and her instructors and classmates said that she wasn't one to lose easily. Sally went on to attend Swarthmore College, where she continued playing Tennis and was almost to become a nationally ranked player. However, after three semesters she dropped everything and headed back to California as she found East Coast winters too harsh. Here she studied physics at the University of California, Los Angeles. Later she joined Stanford University and earned her bachelor's degree in majored in english and physics. She continued to study in the same university completed her masters in physics. She also did her Ph.D. in physics, with research in the field of interaction of X-rays with the interstellar medium. Sally Ride was still in college, finishing her PhD studies, when she came across NASA's advertisement in the Stanford student newspaper seeking applicants for the space program. Sally perfectly fitted the qualifications and conditions required by NASA.

She looked at the advertisement and said "I'm one of those people," she told The New York Times in 1982.

Sally's application was one of the 8,000 people who responded to the advertisement. She was called for the interview and was finally selected in 1978 along with five other women to join the new class of NASA astronauts. Around this time NASA was just about to launch the Space Shuttle, a new technology in space programs. Sally was naturally ecstatic at her achievement but was to face numerous challenges during her training as an astronaut. Most of the earlier astronauts had been former military test pilots and women's physical and mental stamina during stress was always in question by the press and within the organization. One of the trainee astronauts, from her class, was dismissed by Chuck Yeager, the first commandant of the USAF Aerospace Research Pilot School, which produced astronauts for NASA. NASA however had taken cognizance of the gender and minority gap in its space programs that had happened over the years and was keen on changing the environment in the conservative organization.

Although, prior to Sally, the Soviets Union had already sent two women into space, but the public perception as well as 'fellow astronaut,' views had not changed much. In fact, it was reported that as a Soviet woman cosmonaut came onboard the space station, her fellow a male cosmonaut commented that the kitchen and an apron were all ready for her.

Dr. Sally ride was known by her colleagues to keep her cool under stress. The attribute also helped her politely face, the all sorts of questions pertaining to her gender. She was asked—Would spaceflight affect her reproductive organs? Did she cry on the job?

The CBS News reporter Diane Sawyer asked her to demonstrate a newly installed privacy curtain around the shuttle's toilet.

On "The Tonight Show," Johnny Carson joked that the shuttle flight would be delayed because Dr. Sally Ride had to find a purse to match her shoes.

At a NASA news conference, Dr. Ride said: "It's too bad this is such a big deal. It's too bad our society isn't further along." She insisted that she saw herself in only one way—as an astronaut.

At NASA Dr. Sally Ride underwent rigorous training for becoming an astronaut. She trained in water survival tactics, parachute jumping, huge G-forces felt at the time of a rocket launch and weightlessness. She also learned to fly a jet plane. Although, she held a masters in physics, she did well at engineering too and played an important role in the development of the robotic arm, 'Canadarm' for the spaceshuttle. Her expertise at handling the device (Canadarm) scored her over the rest of the class of astronauts and Robert L. Crippen, the commander of the Challenger chose her for the 1983 mission.

Finally on June 18, 1983, she flew aboard the space -shuttle Challenger for STS-7 with four other crew members and wrote her name in history as the first American woman to go in space. The mission lasted about six days, during which Dr. Sally Ride operated Canadarm to deploy and retrieve a satellite.

Some 250,000 odd people crowded at Cape Canaveral to see the launching of the spaceshuttle that Sally rode to space. Many of them wore T-shirts with the words 'Ride, Sally Ride'. The phrase came from the lyrics of the song 'Mustang Sally'..

Gloria Steinem, Ms. Magazine's editor at the time, reported:

> "Millions of little girls are going to sit by their television sets and see they can be astronauts, heroes, explorers and scientists."

In coming years Sally was once more chosen to be the part NASA's next mission in space. In October 1984 she again flew aboard Challenger, this mission lasted for eight days. Sally spent more than 343 hours in space. Sally was on the roster for yet another flight to space, STS-61-M, a TDRS deployment mission, for which she had trained for eight months. However, the mission was cancelled when the space shuttle Challenger blew up on January 28, 1986, 73 seconds after taking off from Cape Canaveral. Instead, now Sally was named as a member of the Rogers Commission. The presidential commission investigated the Challenger accident. Sally Ride headed the commission's subcommittee on operations. As a panel member Dr. Ride asked some very tough questions. It came to light after the panel's investigations that the signs of trouble were already observed on Challenger's earlier flights to space but they deemed as not critical and dismissed by authorities.

Sally Ride was later assigned to NASA headquarters in Washington, D.C. Here she headed a study team that suggested NASA about the future direction of its space program. It was NASA's first strategic planning effort. The report submitted by Sally and her team was titled "NASA Leadership and America's Future in Space." The recommendation in the report included an outpost on the Moon, though not a 'Race to Mars'. The Lunar outpost according to Sally was to be combination of, "Adventure, science, technology and perhaps the seeds of enterprise." The report further stated that Mars should still remain the, 'ultimate objective'. She also said in the report that the US had, 'lost leadership', to the Soviet Union in a number of aspects of space exploration.

Sally Ride left NASA to work at the Stanford University and later at the University of California, San Diego. She also steered two public-outreach programs for NASA—the

ISS EarthKAM and GRAIL MoonKAM projects. EarthKAM lets middle school students take pictures of Earth using a camera on the International Space Station. Students then study the pictures.

Sally was passionately driven to the cause of driving young school children, especially girls towards taking interest in science. In 2001 she started a company, 'Sally Ride Science', to "Make science and engineering cool again." The company creates entertaining science programs and publications for upper elementary and middle school students, with a particular focus on girls.

In 2003, she served on the Columbia Accident Investigation Board. Later in an interview she recounted that part of the problem at NASA was that people had forgotten some of the lessons learned from the Challenger accident. The panel had had already concluded by that time that the disintegration of the shuttle Columbia over Texas was triggered when a chunk of foam insulation fell off the external fuel tank and gashed the leading edge of the wing.

Ride died on July 23, 2012, at the age of 61, seventeen months after being diagnosed with pancreatic cancer.

She was an extremely private person and guarded her personal life fiercely. After her death, her obituary revealed about her same sex partner Tam O'Shaughnessy, a professor emerita of school psychology at San Diego State University and Sally's childhood friend. The relationship was also confirmed by her sister. Dr. Sally Ride is also the first known LGBT astronaut.

Sources

https://en.wikipedia.org/wiki/Sally_Ride
http://www.nasa.gov/audience/forstudents/k-4/stories/

nasa-knows/who-was-sally-ride-k4.html
http://www.biography.com/people/sally-ride-9458284#synopsis
http://io9.gizmodo.com/the-secret-life-of-sally-ride-the-first-american-woman-1586255004
http://edition.cnn.com/2012/07/23/us/sally-ride-dead/
http://www.nytimes.com/2012/07/24/science/space/sally-ride-trailblazing-astronaut-dies-at-61.html?_r=0

□

BUZZ ALDRIN

(The Second Person to Step on Moon)

Buzz Aldrin was born Edwin Eugene Aldrin, Jr. on January 20, 1930. Aldrin is an engineer apart from being an American astronaut. He is most remembered as the second person to step on moon while on the same mission (Apollo 11) as Neil Armstrong. Aldrin had been a fighter pilot and participated in Korean War before joining NASA.

Edwin Eugene Aldrin, Jr. was born to Edwin Eugene Aldrin Sr and Marion (Moon) Gaddys. The Mountainside Hospital in which Aldrin was born spans two across two places Glen Ridge and Montclair, in Essex County, New Jersey. His birthplace is marked as Glen Ridge in the birth certificate. Buzz attended Montclair High School and graduated from it in 1947. Aldrin's maternal grandfather was an Army Chaplain. Edwin Eugene Aldrin Sr, Aldrin's father was in the U.S. Air Force. Aldrin was also a Boy Scout and earned the rank of Tenderfoot Scout.

Edwin Eugene got his nickname Buzz when he was a child. His sister could not say brother and instead pronounced the word as 'buzzer'. Buzzer was shortened by family to Buzz. Aldrin legally adopted this name as his first name in 1988.

He was a good scholar and was offered full scholarship from the Massachusetts Institute of Technology but Aldrin chose to join United States Military Academy at West Point, New York instead, probably inspired by his father to do so, who himself was a career military man. The discipline and strict regimens of the Military Academy went well with Buzz Aldrin and in his freshman year he came in first in his class. He graduated from the Academy in 1951 with a Bachelor of Science in mechanical engineering and was amongst the top three students of his class.

Aldrin's father, who himself was with the U.S. Air Force, suggested his son to continue his studies with the multiengine flight school, so that someday he could command his own crew. However, Buzz dreamt of becoming a fighter pilot. After the summer of 1951, Buzz officially entered the United States Air Force. A bright scholar as he always had been, this time too he was amongst the top students in flight school. Buzz began his training as a fighter pilot later in the year.

Aldrin joined the United States Air Force as a Second Lieutenant and served in 51st Fighter Wing as a jet fighter pilot. He flew F-86 Sabre Jets during the Korean War and was the part of 66 combat missions in the war. Aldrin's wing shot down 61 MiG-15 aircrafts and grounded 57 others in one month of combat while defending South Korea from the invasion of Communist forces in North Korea. Out of the total of 61, Aldrin shot down two MiGs. Later he was decorated with the Distinguished Flying Cross for his service during the war. The June 8, 1953 issue of *Life* magazine carried photos of a Soviet pilot coming out the damaged aircraft; the photos were taken by Buzz Aldrin using his gun camera.

Following the cease fire declaration in 1953 between North Korea and South Korea, Aldrin's next charge was at Nellis Air Force Base, Nevada as an aerial gunnery instructor.

After this he served as an aide to the dean of faculty at the newly open United States Air Force Academy. The same year, in 1955 Aldrin also graduated from the Squadron Officer School located in Maxwell Air Force Base, Alabama. He served as a flight commander at Bitburg Air Base, West Germany and flew F-100 Super Sabres as the part of the 22d Fighter Squadron. He had also returned to school during this time and planned to do masters from Massachusetts Institute of Technology, but he ended up earning a Doctor of Science Degree in Astronautics, in 1963. The subject of his thesis was "Line-of-sight guidance techniques for manned orbital rendezvous".

In the "dedication", of the thesis he wrote:

> "In the hope that this work may in some way contribute to their exploration of space, this is dedicated to the crew members of this country's present and future manned space programs. If only I could join them in their exciting endeavours!"

The work focused upon the study of bringing piloted spacecraft into close proximity with each other.

His desire to become an astronaut was evident in the thesis work, but his first application for astronaut program was rejected because at that time one of the conditions to join the astronaut corps was to be a test pilot. Later that pre-requisite was scrapped and upon his re-applying, Buzz was accepted into the third astronaut class.

The prime crew members of Gemini 9, Elliot See and Charles Bassett lost their life in a tragic aircrash accident. The accident necessitated promotion of Buzz Aldrin and Jim Lovell as the backup crew for the mission. The main objective of this revised mission, now called Gemini 9A, was to rendezvous and dock with a target vehicle. However, the command pilot Thomas P. Stafford and pilot Eugene Cernan were unable to dock with the Augmented Target Docking Adaptor. Next scheduled task was for Cernan to perform a two-hour extravehicular activity, during which he planned to demonstrate free flight in a self-contained rocket pack, the Astronaut Manoeuvering Unit. Again this task could not be accomplished due to stress, fatigue, and overheating.

Aldrin learned from these failures and made Gemini 12, the last Gemini mission, which was also the last chance to prove methods for extravehicular activity (EVA), a success. The Gemini 12 mission flew to space and during its November 11 to November 15, 1966, space flight, Aldrin made a five-hour long spacewalk. He also manually recalculated the docking manoeuvers on the flight, after the failure of onboard radar.

Aldrin's next journey to space got him a special mention in history as the second person to step on moon. He along with Neil Armstrong was chosen for the crew of Apollo 11 and made the first lunar landing on July 20, 1969. His first words on the moon were,

"Beautiful view." Then as Armstrong asked, "Isn't it magnificent?"

He responded, "Magnificent desolation."

His radio message to earth after landing on moon was:

"I'd like to take this opportunity to ask every person listening in, whoever and wherever they may be, to pause for a moment and contemplate the events of the past few hours, and to give thanks in his or her own way."

Aldrin, was a follower of Presbyterian Church, and the

first person to hold a religious ceremony on the Moon. But this communion ceremony was not disclosed to public because NASA did not want negative public criticism again. Madalyn Murray O'Hair, an atheist activist had already entangled NASA in a lawsuit because of the reading of Genesis (from Bible) by the crew members aboard Apollo 18.

Later when the word about the secret communion went around, Aldrin in one of his interviews said:

> "Perhaps, if I had it to do over again, I would not choose to celebrate communion. Although it was a deeply meaningful experience for me, it was a Christian sacrament, and we had come to the Moon in the name of all mankind – be they Christians, Jews, Muslims, animists, agnostics, or atheists. But at the time I could think of no better way to acknowledge the enormity of the Apollo 11 experience than by giving thanks to God."

Although, NASA had issued strict instruction to the crew of Apollo 11 asking them to refrain from making any overt religious comment, Aldrin used a verse from the Psalms to express the magnanimity of his experience. He quoted Psalm 8 (verses 3 and 4):

"When I consider the heavens, the work of thy fingers, the moon and the stars which thou hast ordained, what is man that thou art mindful of him?" However, this quote was passed off as his personal observation and did not create any further controversy for NASA.

Aldrin left NASA in July 1971 and joined Edwards Air Force Base, California as the Commandant of the U.S. Air Force Test Pilot School. Aldrin retired from active duty after 21 years of service and returned to his earlier organization, the US Air Force, in a managerial role.

Aldrin went through a tough phase after NASA, accounts of which he recollected in his auto-bio-graphy, Return to Earth (published in 1973), and memoir, Magnificent Desolation (published in June 2009). During this time Aldrin went through clinical depression, struggled with divorce and alcoholism. He sought treatment for his problems and his resilient spirit helped him take control of his life again.

Aldrin has also received three U.S. patents for his diagrams of a modular space station, Starbooster reusable rockets, and multicrew modules. In 1985, he proposed a special spacecraft system called "Aldrin Mars Cycler," for future missions to Mars. The proposed system claims to use lesser propellant than the conventional fuel. The journey to Mars on the "Aldrin Mars Cycler," is expected to be of five and a half month duration. The design is still in developmental process under the guidance of Aldrin and Purdue University engineers.

Even after retirement from active duty, Aldrin has led an active life. Aldrin and space cannot be isolated ever. In association with the University of North Dakota, founded the Space Studies graduate program. He also produces "Buzz Aldrin's Race into Space", in 1993, which is a computer strategy game. Aldrin also created a rap single and a video called "Rocket Experience", in collaboration with Snoop Dogg, Quincy Jones, Talib Kweli, and Soulja Boy. The song and video was brought out just in time to commemorate the 40th anniversary of the first lunar landing. The proceeds from the sales went to "ShareSpace", the non-profit organization by Aldrin.

Aldrin makes use of all social media to reach his fans to promote the cause of space programs. His views upon space and space programs are always welcome by the public and experts. In an interview with C-SPAN on July 22, 2009 he referred to a monolith on Mars's moon Phobo and mentioned God and Science in one breathe. He said, "We should go boldly where man has not gone before. Fly by the comets, visit asteroids, visit the moon of Mars. There's a monolith there. A very unusual structure on this potato shaped object that goes around Mars once in seven hours." When people find out about that they're going to say "Who put that there? Who put that there?" The universe put it there. If you choose, God put it there...."

Aldrin was seen giving tough competition to young dancers in American reality show "Dancing with the Stars", in 2010. He has also had guest spots on such shows as 30 Rock.

He lent his voice to the parodied versions of himself in two of the animated series by Matt Groening's; Futurama's episode "Cold Warriors" and The Simpsons. In the episode titled "Deep Space Homer" of The Simpsons, his animated self and Homer Simpson makes a trip into space under NASA's plan to improve its public image.

In 2012, Aldrin played himself in a short role in the film Transformers: Dark of the Moon. He meets other characters in the movie Optimus Prime and the Autobots to explain them that during the Apollo 11 mission, the crew found out about a Cybertronian ship already there on the moon, but the discovery was never made public.

Again in 2012, Aldrin made a cameo appearance in Japanese drama film Space Brothers. Aldrin also made an appearance as himself the Big Bang Theory episode, "The Holographic Excitation", where he was seen distributing candies to children on a Halloween night. Aldrin franchise has a long list and also includes a voice-over in a video game, Mass Effect 3 that was released in 2012. Here his character plays a stargazer and appears in the game's final scene.

Aldrin has written several children books also. The list includes *Reaching for the Moon* and *Look to the Stars*; the historical – Men from Earth, and two science-fiction novels, The Return and Encounter with Tiber.

□

CHRIS HADFIELD
(The First Canadian to Walk in Space)

Chris Hadfield was born on August 29, 1959 in Sarnia, Ontario, Canada to Roger and Eleanor Hadfield. The erstwhile astronaut of Canada now lives a retired life. He became a celebrity when he tweeted regularly about his life at the International Space Station during his five months long stay in space. He also holds the record for being the first Canadian to walk in space. Hadfield had been into space three times and was commander of the International Space Station on one of his space missions.

Chris Hadfield wanted to become an astronaut ever since he was a child. At the age nine he watched Apollo 11 Moon landing on TV and was fascinated by the complexities of space. He grew on a corn farm in southern Ontario and went to high school in Oakville and Milton. He married his high-school girlfriend Helene, and they have three adult children: Kyle, Evan and Kristin Hadfield. Chris was a ski instructor at Glen Eden Ski Area before he trained to become a test pilot. He received his license for a glider pilot as a member of the Royal Canadian Air Cadets. He completed his engineering degree from Royal Military College after joining Canadian Armed Forces. He always loved flying and had a chance to fly many types of aircraft in the military. Later he served as a test pilot flying several experimental planes. Hadfield also obtained his master's in aviation systems from the University of Tennessee Space Institute Exchange Program with the United States Navy and United States Air Force.

Chris is a die-hard fan of Toronto Maple Leafs (Canadian professional ice hockey team), and under his space suit wore the team's jersey to space during the re-entry of Soyuz TMA-07M in May 2013. While in space, he tweeted a photo of himself holding a Maple Leafs logo after the end of the 2012 NHL (National Hockey League) Lockout. His statement was "Ready to cheer on from orbit," directed at the team he supported. He also sang the Canadian National Anthem when Toronto Maple Leafs and Montreal Canadians met on field on 18 January 2014.

Hadfield was brilliant scholar from the beginning. Until his senior year he studied in White Oaks Secondary School in Oakville, Ontario. Later he attended Milton District High School and graduated from there as an Ontario Scholar in 1977. At the young age of 15 he had already earned the glider pilot scholarship and in coming year at 16 received powered pilot scholarship.

After high school, he joined the Canadian Armed Forces. For next four years he studied in two different military colleges, for initial two years he attended Royal Roads Military College and for the next two years, the Royal Military College. In 1982 he received his bachelor's degree in mechanical engineering. During the graduation course he also attended CFB Portage la Prairie and completed his Basic Jet flight training from there. In the following year he went Basic Jet Training at CFB Moose Jaw where he received the honours as the top graduate.

Thereafter, he trained at 410 Tactical Fighter

Operational Training Squadron at CFB Cold Lake to become a tactical fighter pilot. During the training he flew Canadair CF-116 Freedom Fighter and the McDonnell Douglas CF-18 Hornet. After the completion of his training he flew CF-18 Hornets as the part intercept missions for NORAD. He intercepted a Soviet Tupolev Tu 95 long-range bomber located in the Canadian Arctic and became the first CF-18 pilot to do so.

Coming few years of Hadfield was spent in Edwards Air Force Base attending the U.S. Air Force Test Pilot School. He also went to Strike Test Directorate in Patuxent River Naval Air Station where he served as an exchange officer with the U.S. Navy from 1989 to 1992. He accomplished various feats one of which was doing research work with NASA on pitch control margin simulation and flight. Hadfield did his masters in aviation systems and received his degree in 1992 from the University of Tennessee Space Institute. His thesis for a master's degree included study of aerodynamics of the F/A-18 Hornet fighter jet during the high-angle attack. Over the years Hadfield has flown more than 70 different types of aircraft.

Hadfield and three other Canadian astronauts were selected in June 1992 to become Canadian astronauts. The selection was tough as the total applications for future astronauts were 5,330. The Canadian Space Agency (CSA) assigned Hadfield to the NASA Johnson Space Center in Houston, Texas in August. Here he was a part of Shuttle Operations Development and look after technical and safety issues. He was also involved in the project developing glass shuttle cockpit. Alan duties at NASA also included assisting shuttle launches at the Kennedy Space Center, Florida. Hadfield, himself flew aboard three space missions and was part of at least 25 space shuttle missions as NASA's Chief CAPCOM. As a capsule communicator (CAPCOM) he was the voice from the ground that coordinated with the astronauts in the space. Hadfield also represented other astronauts from CSA coordinating their activities in the position of Chief Astronaut for the CSA.

Hadfield's first space mission STS-74, came in November 1995 when he flew aboard Atlantis in the capacity of Mission Specialist 1. This was the second mission from NASA that involved docking of its space -shuttle with Mir, the Russian Space Station. Hadfield and other crew members of the spaceshuttle Atlantis transferred various scientific supplies along with food and water that weighed more than 1,000 kg in all. This mission marked not only the entry of the first Canadian into space but already trained and prepared Chris Hadfield was first from his country to operate the Canadarm in orbit, and the only Canadian to board Mir ever.

Hadfield's second mission (STS-100) to space in April 2001 made him set yet another record of being the first Canadian astronaut to walk in space. In this mission again Hadfield was Mission Specialist 1 on International Space Station (ISS) assembly Flight 6A. Chris and his fellow crew members aboard Space Shuttle Endeavour installed Canadarm2, the Canadian-made robotic arm along with Raffaello, the Italian-built resupply module. The mission lasted eleven days and Chris Hadfield walked in space twice during this time. His first spacewalk was not entirely trouble-free. Hadfield felt severe eye irritation as he stepped out of the spaceshuttle, the sensation almost blinded him temporarily and he had to vent oxygen in pace to counter the effect. The reason for the condition was attributed to the anti-fog solution used to polish the spacesuit visor. But this did not deter him from spacewalking again during the same mission. He clocked 14 hours, 50 minutes in his spacewalks and travelled 10 times around the world while spacewalking.

Hadfield embarked on his third and perhaps the most adventurous of his space missions on 19 December 2012. He flew aboard Soyuz TMA-07M to reach the International Space Station (ISS) for a five month long stay there. Hadfield, with the enthusiasm of any ordinary youngster, was eager to make this expedition a success. Shortly before his departure to space as part of Expedition 35 he said:

> "To be able to command the space station, yes, it's professional, and yes, I'll take it seriously, and yes, it's important for Canada, but for me, as just a Canadian kid, it makes me want to shout and laugh and do cartwheels."

This mission also holds various other firsts for Chris Hadfield individually and for Canadian astronauts. He commanded the ISS for three months from March 2013 (when Expedition 34 crew went back) to 12 May 2013 (until Hadfield returned to the earth) and became the first Canadian to do so.

"Good morning, Earth." Hadfield would tweet every morning during his stay at ISS. His another 'first' include using social media profusely to reach out to people on earth making them accessible place. He was widely written and talked about in print and on T.V. when he, with the help of his tech savvy son, uploaded a video on YouTube of his rendition of the British singer David Bowie's song 'Space Oddity'. The video soon went viral receiving more than 7 million views within a few days of uploading. Chris returned safely to the earth on May 13, 2013 with his fellow astronauts Roman Romanenko from Russia and Tom Marshburn from America. After his long stay in space he was reported saying:

"It smelled of just wind in the grass," upon being asked how it felt when he came out of the spacecraft after landing.

As of March 2016 figures Hadfield has over 1,560,000 Twitter followers. *Forbes* magazine described Hadfield as "Perhaps the most social media savvy astronaut ever to leave Earth." He also runs a popular blog on Tumblr. His online conversations with various Star Trek actors including William Shatner caught media attention widely. Hadfield actually lived the fictional adventure that Star Trek franchise has created.

Apart from his famous firsts achievements during space missions Hadfield also served as the Commander of NEEMO 14 from May 10 to 23 in 2010. NEEMO 14 (NASA Extreme Environment Mission Operations) was an undersea analog mission. The mission included simulation of various activities, which astronauts likely to perform during space missions. The underwater facility is located off the Florida coast. NEEMO 14 activities used the ocean floor as a simulated surface where exploration missions could be conducted as they would be carried out on the surfaces of asteroids, moons and Mars. These experiments gave better perspective of the astronauts' interaction with space equipment such as advanced spacesuits, a rover, a lander, and robotic arms.

In June 2010, Hadfield was part of the Pavilion Lake research team, located 420 km northeast of Vancouver. Pavilion Lake is one of the few places on Earth where microbialites are found. The team used a combination of remotely operated vehicles, autonomous underwater vehicles, SCUBA divers and DeepWorker submersibles to help understand how the microbialites formed and possibly make it easier to identify potential forms of extraterrestrial life on future missions to Mars.

Hadfield was also an important part of the Pavilion Lake research team. Pavilion Lake is located 420 km northeast of Vancouver. The place is amongst very few sites where microbialites are found. The research team studied the formation of microbialites with the help of SCUBA divers, DeepWorker submersibles, and SCUBA divers. The research finding would help in identifying potential forms of extraterrestrial life on missions to Mars.

Hadfield has donned various hats during career with military and NASA, one of which is of an amateur singer. While living aboard ISS during Expedition 35, he found out time to record music using the Larrivée Parlor guitar brought earlier to the space station. The first song he recorded in space was Jewel in the Night, which was hosted on YouTube. Hadfield had credited musically on various albums including that of his brother Dave Hadfield, with whom he also performed the 'Canada Song', released later on Canada Day in 2014. Chris Hadfield released Space Sessions: Songs From a Tin Can, in October 2015, an album with collections of songs he recorded while on the International Space Station. His version of 'Space Oddity' by David Bowie received 28 million views on YouTube as per the January 2016 (shortly after Bowie's death) figures. The YouTube video was also discussed in media for the legal implications related to the copyright issues when someone publicly performs a work copyrighted on earth while in orbit.

Hadfield announced his retirement on June 2013, and was relieved of his duties from the CSA on July 3, 2013. After his active and adventurous tenure with the NASA, CSA, and military, Hadfield continues to enthrall his followers. He shared his experiences and excitement with space lovers in his memoir, An Astronaut's Guide to Life on Earth, released in 2014. The book was a New York Times best-seller. In his book he says:

> "There are no wishy-washy astronauts. You don't get up there by being uncaring and blasé. And whatever gave you the sense of tenacity and purpose to get that far in life is absolutely reaffirmed and deepened by the experience itself."

Hadfield also loves photography and clicked many pictures while on his three space missions. He published a collection of these photos in 2014, in an album titled: You Are Here: Around the World in 92 Minutes.

Retired from his 35 years long career as an astronaut and a military pilot Hadfield continues to be YouTube sensation and an engaging speaker through TED talks.

Sources

https://en.wikipedia.org/wiki/Chris_Hadfield
http://chrishadfield.ca/about/
http://www.asc-csa.gc.ca/eng/astronauts/biohadfield.asp
https://www.ted.com/talks/chris_hadfield_what_i_learned_from_going_blind_in_space?language=en
http://www.biography.com/people/chris-hadfield-21230027#synopsis

□

KALPANA CHAWLA
(The First Indian-born Woman in Space)

Kalpana Chawla was an Indian-American Astronaut and was the first woman of Indian origin in space and second Indian in space after Rakesh Sharma in 1984. She flew to space twice, but unfortunately her second flight could not land on successfully and she was killed along with seven members in the Space Shuttle Columbia disaster on February 1, 2003. In her course of two flights she logged 30 days, 14 hours and 54 minutes in space. She was an ordinary girl with the dreams of touching the sky. She with her efforts and focus on her aim reached the stars.

Kalpana was born on July 1, 1961, in small town in Karnal in Haryana, India. Her father's name was Banarsi Lal Chawla, who was a leading industrialist of Karnal and owned a tire factory, her mother Sanyogita was a housewife. Kalpana was the youngest of the four siblings, she had two elder sister named Sunita and Deepa, and a brother named Sanjay. Her family was expecting a boy as the last child but instead a girl-Kalpana was born. Nevertheless, she was the most pampered child. Her family was refugee Hindu migrants from Pakistan and they settled in Karnal after the partition of India and Pakistan in 1947. Despite much hardship, her family managed to provide good education to Kalpana and her siblings. Kalpana's elder sister and her sister-in-law are ardent bird watchers in Delhi and her brother is a businessman. Her early education started in India; she did her schooling from Tagore Public School.

In childhood she and her brother used to share interest in flying and she was hugely inspired by J.R.D. Tata, who was India's first pilot. Her mother once mentioned in an interview that "Kalpana was a very different child from her childhood, She used to cut her own hair, never wore ironed clothes, learned karate." In a male dominated society, she never gave up her dream of flying just because of a girl and to fulfil her ambition she took her Bachelor of Science degree in aeronautical engineering from Punjab Engineering College, India, where she was the first woman in her batch. After her Bachelor degree, she moved to United States for obtaining her Master of Science degree in aerospace engineering from University of Texas at Arlington in 1984.

She met her future husband Jean Pierre Harrison on the day, she landed in U.S.A., who was a flying instructor. Harrison was attracted towards her the moment he saw her the first time. Later he introduced Kalpana to flying, hiking, back-packing, and reading. Kalpana's brother helped persuade their parents to let his sister marry Harrison they were married on December 2, 1983 and she became naturalised United States citizen in 1990.

After her death he wrote a book titled "The Edge of Time: The Authoritative Biography of Kalpana Chawla" which got published in 2011. He later re-married and had a son.

Kalpana was a certified flight instructor whose work was to rate aircrafts and gilders. She also had the commercial pilot license for single and multiengine airplanes, hydroplanes and gliders. Kalpana was a licensed

Technician Class Amateur Radio Person certified by the Federal Communication Commission.

In 1988, she completed her Doctorate from University of Colorado. During the tome while pursuing her Ph.D. she started working at the NASA Ames Research Center in the San Francisco Bay area, where she did research on "The simulation of complex airflows encountered around spacecraft."

Afterwards, Kalpana started working with Overset Methods, Inc. in Silicon Valley. She also served as the vice president and as a research scientist specializing in simulation of moving multiple body problems. After becoming a naturalized U.S. citizen in 1991, she then applied for the NASA Astronaut Corp (a unit of the United States National Aeronautics and Space Administration) (NASA) that selects, trains, and provides astronauts as crew members for the U.S. and international space missions. The unit is based at Lyndon B. Johnson Space Center in Houston, Texas. In March 1995, Kalpana joined the corp and was also selected for her first mission to travel to the space in 1996. After her selection, Kalpana reported to the Johnson Space Center in March 1995 as an astronaut candidate in the 15th Group of Astronauts for a yearlong training and evaluation. There she was assigned as crew representative to work on technical issues for the Astronaut Office EVA/Robotics and Computer Branches.

Her first space mission started on November 19, 1997 in STS-87, Columbia. She made history by being the first woman of Indian origin in space. On this mission Kalpana was accompanied by six other crew members. While travelling in the space she said "You are just your intelligence",

Kalpana travelled 10.67 million km and about 252 times

around the Earth. She was assigned as mission specialist and prime robotic arm operator on STS-87; she was also responsible for deploying the Spartan Satellite, which malfunctioned necessitating a spacewalk by Winston Scott and Takao Doi to capture the satellite. The mission ended on December 5, 1997, logging more than 372 hours in space. The STS-87 was focused on experiments designed to study how the weightless environment of space affects various physical processes and also to study the various atmospheric outer layer of the sun. After the successful completion of STS-87 flight, Kalpana was given various technical positions at the astronaut office to work on the space station. Her excellent performance was recognized and rewarded.

After her first launch, she had said, "When you look at the stars and the galaxy, you feel that you are not just from any particular piece of land, but from the solar system."

She was again selected for her second flight STS-107 in 2000. She was given the responsibility to perform microgravity experiments which included almost 80 experiments. Along with her team members, she undertook a detailed training and research on advanced technology development, astronaut health and safety and the study of Earth and space science, due to one or the other reasons this mission was repeatedly rescheduled various times, once due to technical problems, another time because of cracks in the shuttle engine flow liners and other faults which delayed the mission for long. Finally the delays ended in 2003 and the flight was launched on 16 January 2003 from the Kennedy Space Center – Florida, the flight stayed in space for 16 days and the crew conducted various experiments. This was Columbia's (name of the shuttle) 28th spaceflight and the shuttle was believed to be good for 100 flights. The STS-107 mission came to a tragic end, when after re-entering the earth's atmosphere the Space Shuttle disintegrated over Texas, US, just 16 minutes before its landing on the earth. All the seven crew members; Commander: Rick D. Husband, Pilot: William C. McCool, Payload Commander: Michael P. Anderson, Payload Specialist: Ilan Ramon, Mission Specialist: David M. Brown, Mission Specialist: Laurel Blair Salton Clark, of the STS-107 mission were killed in this unfortunate event, including the Mission Specialist and arm operator Kalpana Chawla and the first woman of Indian origin astronaut.

It is believed that the failure of STS-107 was caused by a piece of foam that broke off during launch and damaged the thermal protection system (reinforced carbon-carbon panels and thermal protection tiles) on the leading edge of the orbiter's left wing. During re-entry the damaged wing slowly overheated and came apart, eventually leading to the loss of control and disintegration of the vehicle.

When the news of the STS-107 crash was flashed on television, millions of people around the world and especially the people of Karnal were praying and hoping that "May Kalpana and the crew manage to survive somehow."

A journalist wrote about her, "She was a heroine. It takes enormous ability to become an astronaut. You need to know a lot about everything from biology to astrophysics to aeronautical engineering. In this age of super-specialization, you must have encyclopedic knowledge to be an astronaut. Her achievement is aweinspiring."

Her achievements inspires others to keep working towards success. In an interview she had said, "You couldn't lose by working hard and everyone seemed to follow that rule, it helped instill the notion that no matter what the circumstances, you could indeed follow your dreams."

Kalpana had received many awards during her life as well as after her death; some of them are the Congressional Space Medal of Honour, NASA Space Flight Medal and the NASA Distinguished Service Medal received by her during her lifetime. Many awards and memorials have been instituted in the honour of Kalpana after her death, The Outstanding Recent Alumni Award at the University of Colorado, given since 1983, was renamed after Kalpana Chawla. In Karnal about 30,000 students and citizen joined hand to make a humanchain of about 36.4 km long to support the demand for a Kalpana Chawla medical college in the city of Karnal. The gesture showed the magnitude of respect and love, people had in their heart for her. The Haryana Government thus accepted this long pending demand of people and a Kalpana Chawla Medical College has been set-up at Model Town, Haryana. After her last and unfortunate mission, India has renamed its first weather satellite 'Kalpana-1' in her honour. Many scholarships are being given in her name. The Kalpana Chawla ISU Scholarship fund was found by alumni of the International Space University in 2001. The scholarship's objective is to support Indian student participation in international space education programs, The Kalpana Chawla Memorial Scholarship Program was instituted by the Indian Students Association (ISA) at the University of Texas at El Paso. The Indian Institute of Technology, Kharagpur, named the Kalpana Chawla Space Technology Cell in her honour. The 74th Street in Jackson Heights, Queens, New York City has been renamed Kalpana Chawla Way in her honor. A Planetarium was also made by Haryana Government in her honour at Jyotisar, Kurukshetra, which was named Kalpana Chawla Planetarium. A song called 'Contact Lost', has also been created by Steve Morse from the band Deep Purple in the memory of the Columbia tragedy. Her honours and achievements are uncountable; her contribution to the Space can't be measure in any terms.

Remembering Kalpana, her brother Sanjay said, "To me, my sister is not dead. She is immortal. Isn't that what a star is? She is a permanent star in the sky. She will always be up there where she belongs."

Kalpana Chawla died a hero. She has been a great inspiration and motivation to people around the world, especially to girls from small towns, who thought they can never fly in the sky.

Kalpana once said, "The path from dreams to success does exist. May you have the vision to find it, the courage to get on to it, and the perseverance to follow it."

The story of Kalpana Chawla – a girl from a small town who touched the skies had become an inspiration; many people now want their girls to follow the footprints of Kalpana. When asked about how she felt about being a woman in her field she replied, "I really never, ever thought, while pursuing my studies or doing anything else that I was a woman or a person from a small city or a different country. I pretty much had my dreams like anyone else and I followed them. And people who were around me, fortunately, always encouraged me and said, "If that's what you want to do, carry on."

Kalpana was a strict vegetarian, on her mission to space as a part of worldwide campaign she carried a white silk banner to honour teachers, she also carried dozens of CDs of Ravi Shankar, CDs by Deep Purple and other, as she was quiet interested in music.

Despite living in America, she was truly the pride of

India. She frequently visited Karnal to motivate the girls to follow her. There are many books written on Kalpana Chawla's life: 'Kalpana Chawla: A Life' by Anil Padmanabhan, 'Life and Dreams of Kalpana Chawla' by Gurdeep Pandher.

Kalpana Chawla is such a figure who, with her strong will power, hard work and her dedication, rose to such a height. Not only India but the whole world is proud of her achievements.

□

RAKESH SHARMA

(The First and Only Indian in Space)

Rakesh Sharma became the first Indian astronaut to travel to space aboard Soyuz T-11 on 2 April, 1984, Rakesh Sharma. An alumnus of the National Defence Academy, became part of Air Force Cadet in 1966 and joined Indian Air Force as a test pilot in 1970. Sharma's dedication and excellence in his workfield gave him an opportunity of becoming a cosmonaut. He, along with Wing Commander Ravish Malhotra, was selected for the training to go into space under the joint program of Soviet Intercosmos Space Program and Indian Space Research Organisation (ISRO). After a long and demanding training schedule, finally Sharma became the chosen one who would always hold the honour of becoming the first and only Indian citizen so far to travel into space. (Recent years' astronauts Sunita Williams and late Kalpana Chawla are US citizens of Indian origin). Sharma spent around eight days in the outer space. He flew aboard Soyuz T-11, which docked and transferred its crew of three to Salyut 7, where the team carried out various scientific and technical experiments. He was awarded with Hero of Soviet Union, the highest honour in the Soviet Union. The Indian government also conferred on him the Ashok Chakra, the highest gallantry award during peace time. He retired from Indian Air Force in 1987 and joined Hindustan Aeronautics Limited and served as Chief Test Pilot at Nashik and Bangalore until his retirement from test flying in 2001. His words from space, *'Saare Jahan Se Achcha'*, to the then Prime Minister of India Mrs. Indira Gandhi upon being asked how India appears led from space, made every Indian proud at that moment.

Sharma was born on 13 January, 1949 in district Patiala in Punjab. His father was a banker with Punjab National Bank. The job took Sharma's father to Hyderabad later, where Rakesh was actually raised and completed his early education. St. George's Grammar School in Hyderabad. Since at the age of six Sharma had set his heart upon being a pilot, He told "The Hindu", in an interview. He wanted is be a pilot, "Since I was six years, old, and a cousin in the Air Force took me around and showed me airplanes, their cockpits, etc. If you end up doing what you are passionate about, the journey is so easy." He also recollects his memories as a child when he would run down the streets with his arms stretched out pretending to be an airplane.

To realize his childhood dream, after completing his school education, Sharma went on to join National Defence Academy in 1966 as an Air Force Cadet. He was commissioned as a test pilot in Indian Air Force in 1970. As a pilot he got to fly HT2 propeller tail wheel airplane, T6 G, Harvard and later the Vampire airplanes. Just two months later he was flying planes in a war. Sharma was always a dedicated pilot and he got to fly 21 combat missions during war against Pakistan in 1971. During these missions he flew Mikoyan-Gurevich aircrafts, popularly known as MiG-21. He quickly scaled up the ladder in Indian Air Force and was soon appointed as the Squadron Leader and pilot of the Indian Air Force. He learned quickly and soon upgraded his skills to become an experimental test pilot which meant that now he could test designs of the airplanes off the drawing board.

Sharma's dedication and excellence at his work did not go unnoticed and he was chosen to be the part of a mission of national importance. The mission was a combined space program between the two space agencies of India and the Soviet Union viz. Indian Space Research Organisation (ISRO) and the Soviet Intercosmos respectively. The Soviet Union extended an offer to include an Indian along with Russian cosmonauts in their forthcoming space mission to the then Indian Prime Minister Mrs. Indira Gandhi. ISRO declined the offer initially (only to accept it later) citing the reason that the country was not prepared for such programs and it would be in the nation's interest if ISRO is allowed to focus completely upon socio-economic benefits of the ongoing satellite programs. Given the cold war ensuing at that time between the Soviets and the American political counterparts, the offer had a partisan undertone of Soviets trying to win the support of various nations across the Globe. The erstwhile Soviet Union wanted to assert its communist way of life over the American democracy by extending such offers to its allied nations. The motivation behind the project was for India to align itself with The USSR or at least prevent India from standing beside the USA. India so far had collaborated with the USA in the ongoing space race. The first Indian rocket (Nike-Apache India) launched from India into space in 1963 came from NASA. In 1975, India, not so a rich country, could enjoy satellite transmission. The satellite signals were beamed to the interiors of India from American Communication

Satellite, ATS-6. Nevertheless, the proposal if accepted would help India to take a step further in its space program and help India to showcase its technological advancements in front of the rest of the world.

Finally, the Indian government agreed to send its best Air Force Officers to Moscow to be trained as cosmonauts for the upcoming Soyuz T-11 mission. In the beginning the entire project was kept under wraps and the selection process was also a hush-hush program. The code name of the project was Project Pawan. Indian Air Force pilots were just asked to volunteer for something extraordinary mission and Sharma applied along with many other applicants. The list was further pruned and finally now only 20 remained.

The Russian officials and the Indian Air Force shortlisted the names of four officers keeping in mind various parameters including health, experience, merit and finally four cosmonaut aspirants were flown to Moscow. These aspirants had to take some medical test; the equipment required for some which tests were not available in India.

Finally Squadron Leader Rakesh Sharma and Wing Commander Ravish Malhotra were chosen to undergo the cosmonaut training in Moscow. It was not until very last few days before flying off to space the name of the chosen Indian cosmonaut was disclosed. These ace test-pilots of the Indian Air Force were put through a rigorous selection and training process, both in the country and in the erstwhile Soviet Union. The training period lasted 18 months, "Or more importantly two winters," as Sharma observed in his interview later. According to him they had to walk through snow daily for 7-8 km to attend classes. Sharma and Malhotra not used to of such weather conditions were motivated by the fact that everybody in the Star City (an exclusive facility for cosmonauts' residence and training) including the highest of the officials were doing so. The first three months of the training were devoted to learn the Russian language to prepare the Indian cosmonaut for interaction with their Russian counterparts and to enable them to read the labels and instruction which would be written in the Russian language inside the spacecraft. During their yearlong training both the Indian Air Force officers underwent demanding training and also learnt the Russian language.

The other ground classes included studying space flight dynamics, learning about the star sky, simulator work and microgravity. Specialized models of Ilyushin Il-76 were used for zero-G training. The aircrafts were specially stressed and padded from inside and performed a preprogrammed manoeuver at high speed to create the conditions of zero gravity for 25 seconds. Sharma, a test pilot was already used to flying high speed airplanes and the experience came in handy while preparing for space -craft flight. He said later that he never felt air sickness or was uncomfortable during the microgravity training on ground. He reported later that when in space since the exposure to microgravity was for long period, the effects felt on the human body were completely different.

On April 3, 1984, Sharma and two other Soviet cosmonauts, Commander Yury Malyshev and Flight Engineer Gennady Strekalov, launched off from Baikanour cosmodrome in the tiny Soyuz T-11 capsule. The cosmodrome or the launch pad in Baikanour (now in Kazakhstan) is the world's first and largest operational space launch facility. The capsule later docked with the space station Salyut-7. Sharma spent the next seven days,

21 hours and 40 minutes, onboard the space station, Salyut-7. He was the 128th human to go into space, and the only Indian so far with space experience.

Some 26 years later in an interview with *The Hindu*, Sharma mentioned that "Every minute in space was accounted for," and in lighter vein added he wished that he had more time to peer out of the window into space. He also said that he would love to go to the space again but now as a passenger and not a crew member.

In his recollections later about the launching of the spacecraft, Sharma said that the moment can be unnerving as being a test pilot one is used to handling the aircraft himself whereas due to the high acceleration of a spacecraft everything has to be automated and controlled from the ground. All that the riding astronaut has to do during the launch is to try not to move so as to avoid the weight shift.

While aboard Salyut-7, the space station, Sharma carried out several technical and scientific studies including 43 experimental sessions. His responsibilities also included capturing pictures using multispectral camera, especially of the Northern part of India. His work was mainly in the fields of biomedicine and remote sensing.The Images became part of the government data- base to be later utilized for various purposes including identifying glacial shifts or for zeroing in on a location to construct a hydroelectric power station close to the Himalayas. He recounted later that the feeling of being in space as the first Indian astronaut was the best experience ever. He also carried out various experiments designed by Indian scientists using the equipment which were manufactured in India.

The experiments conducted by him also involved materials processing experiments, including silicium fusing tests. He also experimented with practicing yoga to deal with the effects of prolonged orbital spaceflight, and this particular effort was highly appreciated by Roscosmos.

A joint television conference was held with the officials in Moscow and Mrs. Indira Gandhi he remarked are that the geographical boundaries are man-made which not actually visibly demarcated on the earth are. He still holds on to this view and often repeats in his various interviews that the work in outer space should be a collaborative effort. According to him human race should not carry conflicts with themselves to other planets; or the conquest will lead to making of another Antarctica where although no human inhabitation is not there but countries have demarcated their boundaries there too.

Upon his return to the earth, he was celebrated as a hero and honoured with various awards and felicitations. He is also associated with the Indian space programs in advisory position. Sharma was also greeted by some great Russian cosmonauts of his time including Alexei Leonov and Valentina Tereshkova. Tereshkova presented Rakesh with a Samovar, a traditional Russian utensil.

Rakesh Sharma an Indian hero with whom India stepped into the outer space shies away from the limelight and often says that after all these years younger generation often mistakes him as the man who went to the moon. In his words, "It just proves my point that it makes more sense to move on rather than dwell on something that happened so many years ago."

□

SUNITA WILLIAMS

[The Woman Astronaut with the Record of Maximum Number of Space Walks (the Space Walker)]

Born on September 19, 1965 at Euclid-Ohio in America to Dr. Deepak Pandya and Bonnie. Now fifty years old, Sunita is the youngest of three siblings, she has an elder brother and sister. Her full name is Sunita Lyn Williams, Her father was a famous neuroscientist and mother technician. Sunita is an American woman of Indian origin. Being the daughter of a Hindu father and a Christian mother, she has an affinity for both faiths and is known to be a Ganesh devotee. Her roots on her father's side go back to Jhulasan, Mehsana district in Gujarat, India and her mother was a Slovenian. She is married to Michael J. Williams, a Federal Police Officer in Oregon. They have no children but Sunita someday wants to adopt a girl child, girl from Ahmedabad. She is a great pet lover too – her pet Jack Russell Terrier is named 'Gorby'.

Sunita was hardworking right from her childhood. She graduated from Needham High School in Needham, Massachusetts, in 1983. She received a Bachelor of Science degree in Physical Science from the United States Naval Academy in 1987, and a Master of Science degree in Engineering Management from Florida Institute of Technology in 1995. She grew up wishing to be a professional swimmer. At some point in life she also wanted to train as vet.

Sunita joined the United States Navy in May 1987 and was commissioned an ensign. After going through the ranks of Basic Diving Officer and Naval Aviator in July 1989, in 1993, she was assigned to Helicopter Combat Support Squadron 8 (HC-8) in Norfolk, Virginia. Here she participated in two overseas deployments Operation Desert Shield and Operation Provide Comfort and flew over the Mediterranean, Red Sea and the Persian Gulf. She also participated in the relief operations for Hurricane Andrew in September 1992 for which she flew to Miami, Florida in the capacity of Officer-in-Charge of an H-46 detachment. In January 1993, Sunita joined the U.S. Naval Test Pilot School from which she graduated in December. Later she was assigned as the squadron Safety Officer and flew test flights. She returned to the Naval Test Pilot School and served as an instructor in the Rotary Wing Department and as the school's Safety Officer. Sunita had already flown more than 30 different aircrafts and logged more than 2,770 flight hours but destiny had decided something greater than that for her. She applied for astronaut training in 1998 and was selected by American Space Agency NASA in June 1998. Her training started the same year in August. She travelled to Moscow, where she received training in robotics and other International Space Station (ISS) operational technologies while working with the Russian Federal Space Agency and with crews preparing for expeditions to the ISS. During the training she developed the skills in technical briefings, physiological training and preparing for T-38 flight training, and she also learned water and wilderness survival techniques.

She also worked with in the Robotic branch on the ISS Robotic Arm and the related Special Purpose Dexterous Manipulator. She was a crew member on the NEEMO 2 (NASA Extreme Environment Mission Operations), mission, living underwater in the Aquarius habitat for nine days in May 2002.

Sunita was a mission specialist on STS 116. Name of Sunita's shuttle was 'Discovery'. It was launched on December 9, 2006 to join the expedition 14 crew. As a member of Expedition 14 crews she served as a Flight Engineer. In April the Russian members of the crew rotated, changing to Expedition 15. Sunita cut her pony-tail and decided to donate it to Locks of Love (a non-profit organization that makes haitr prosthetics for ill children). Her hair was cut by Joan Higginbotham in the International Space Station. The crew of STS-116 brought the ponytail back to earth.

Sunita's space stay was full of unprecedented incidents. She requested for spicier food while in space. In response to her wish a tube of wasabi was also included in the supplies sent aboard Progress Spacecraft resupply mission. The wasabi tube was packaged at one atmospheric pressure and as Sunita opened the tube in the low pressure atmosphere of the ISS, the gel like paste came out all at once and the spice filled jet became difficult to contain in the no gravity environment. Cleaning the spilled food was nothing less than a challenge afterwards.

"We finally got the wasabi smell out after it was flying around everywhere." We cleaned it off the walls a little bit."

Sunita told her mother later in a conversation from space in the same week after the spill incident. The conversation was arranged by Boston radio station WBZ.

In another such unexpected incident, Sunita's camera came unhooked as she was performing planned extravehicular activity. The camera became untethered may be because of a failed attaching device even before Sunita could react the camera flew off in far space. The debris analysts at NASA studied the camera's trajectory later.

Sunita spent 6 hours and 40 minutes in her third spacewalk and 29 hours and 17 minutes till her fourth spacewalk breaking the record held by Kathryn C. Thornton for most spacewalk time by a woman. On January 31, February 4 and 9, 2007 she completed three spacewalks from the ISS with Michael Lopez Algeria.

Sunita set another 'first', by running the first marathon by an astronaut in space station. On April 16, 2007, she ran the Boston Marathon in space on the station's treadmill with the timing of four hours and 24 minutes. She circled the earth almost three times during this pursuit. Her fellow astronauts cheered her and kept her hydrated with oranges. Sunita's sister Dina Pandya and astronaut Karen L. Nyberg also ran at the same marathon but on the earth. William followed their progress through Mission Control. Sunita again participated in the Boston Marathon in 2008, this time she ran with other participants on the earth.

STS-117 mission aboard 'Atlantis', was sent to the space to bring Sunita back to the Earth on 26th April, 2007. The spaceshuttle Atlantis carrying Williams touched down at the Edwards Air Force Base in California on June 22, 2007. Indian-American astronaut Sunita came back to earth after setting a record of 195 days stay in space. Full of challenges and adventures as her stay in space, the landing of her space shuttle was also not easy. On the day of her scheduled landing Kennedy Space Center in Cape Canaveral had poor weather, causing three failed landing attempts over 24 hours. Mission managers then diverted Atlantis to the Edwards Air Force Base in California, where she finally touched earth.

Sunita flew to space for the second time as a part of Expedition 32/33 on July 15, 2012. She was launched from the Baikonur Cosmodrome aboard Russian spacecraft Soyuz TMA-05M. The spacecraft docked with the International space station on July 17, 2012 at the orbiting outpost. This time Sunita had to stay in space for four months. She was accompanied by Russian cosmonaut Yuri Malenchenko and apan Aerospace Exploration Agency (JAXA) astronaut Aki Hoshide on this mission. Sunita served as a commander of ISS during the space mission Expedition 33. She was the second woman to achieve this feat. During her earlier space mission Sunita ran a marathon and this mission she came prepared to participate in triathlon while in space, The triathlon coincided with the Nautica Malibu Triathlon held in Southern California on the earth. Sunita ran on International Space Station's own treadmill and stationary bike, did the swimming simulation using the Advanced Resistive Exercise Device (ARED). She did weightlifting and resistance exercises that were equivalent to swimming in microgravity. After 'swimming' half a mile (0.8 km), biking 18 miles (29 km), and running 4 miles (6.4 km). Williams finished with a time of one hour, 48 minutes and 33 seconds.

She returned to earth on November 19, 2012. When her spaceshuttle touched down in Arkalyk, Kazakhstan. Helicopters were sent to join the search-and-recovery crew to find and assist the astronauts. The rescue mission was urgently dispatched as the capsule parachuted down some 35 kilometres from the planned touchdown site due to a procedural delay.

She is the second woman of Indian origin to have been selected by NASA for a space mission after Kalpana Chawla. Indian-American astronaut Sunita, who holds the feat of completing the longest space flight for a woman, has also received the Padma Bhushan Award, India's third highest civilian award in 2008. Other than Padma Bhushan she has a long list of awards in her name: Humanitarian Service Medal, The Army Commendation Medal, Achievement Medal, Medal "For Merit in Space Exploration", for her outstanding contribution to the development of international cooperation in manned space flight and she is also the first Non-Resident Indian to be honoured with Sardar Vallabhbhai Patel Vishwa Pratibha Award by World Gujarati Society.

Her recreational interests include running, swimming, biking, triathlons, windsurfing, snowboarding and bow hunting. Sunita is also well connected to Indian culture and has always shown her respect towards it, she took a copy of *Bhagavad Gita*, a small figurine of Ganesha with her to the International Space Station (ISS). In 2015 Sunita was selected as one of four astronauts to make the first test flights in NASA's Commercial Crew program along with Robert Behnken, Eric Boe and Douglas Hurley in which two new private crewed spacecraft, SpaceX's Dragon and Boeing's CST-100, would take astronauts and supplies to the ISS. Flights were scheduled to begin in 2017.

NASA Administrator Charlie Bolden said in a blog post that announced the astronaut selection.

"These distinguished, veteran astronauts are blazing a

new trail, a trail that will one day land them in the history books and Americans on the surface of Mars."

John Holdren, assistant to the President Obama for science and technology, observed:

"Their selection allows NASA to move forward with the training necessary to deliver on President Barack Obama's ambitious plan for returning the launch of the US astronauts to US soil."

Every Indian should be proud of Sunita, she is a great inspiration to not only Indian but to the whole world. She has shown that being a woman cannot become an obstacle in the path of your success, if you are determined you can even touch the sky. We can learn number of lessons from Sunita's life, her career as a navy officer and astronaut was not easy but her attitude made the tough jobs easy for her. She shows that positive attitude, continuous hard work, dedication, perseverance and struggle will surely lead you to success. She is also working for Women's Empowerment, encouraging them to come forward and to make their own individual identity. Sunita spent a total of 322 days in space on two missions and ranks sixth on the all-time U.S. endurance list and second all-time for a female astronaut. With 50 hours and 40 minutes, she also holds the record for total cumulative spacewalk time by a female astronaut.

Sunita is the recipient of Padma Bhushan, the third highest civilian award by Government of India. During her 2007 visit of India, she went to her ancestral village Jhulasan and the Sabarmati Ashram in Gujarat. The World Gujarati Society honoured her with the Sardar Vallabhbhai Patel Vishwa Pratibha Award. This added another 'first' to her already long list of firsts. Sunita the first person of Indian descent who was not an Indian citizen to be presented the award. On October 4, 2007, She spoke at the American Embassy School, and met Manmohan Singh, the then Prime Minister of India.

> "Don't get bogged down by the notion of limits. There aren't any."
>
> —*Sunita Williams*

□

ALAN B. SHEPARD

(The Astronaut Who Played Golf in Space)

Alan Shepard was the first American and second person to travel to space in 1961. He was also the fifth man to walk on the moon. He is remembered as the astronaut who played golf in space. Alan Shepard in his lifetime was regarded as an, intense, colourful, and dramatic person. Before Allan Shepard joined NASA, he was a navalofficer , an aviator and, a test pilot. Later in life he became a businessman as well.

Shepard was born on November 18, 1923 in Derry, New Hampshire, United States, a small village which is few miles south of Manchester. His full name was Alan Bartlett Shepard Jr. He was the eldest of two siblings. His younger sister name was Pauline. His father Lt. Col. Alan B. Shepard Sr. served in the National Guard and then the Army Reserves. His mother, Pauline Renza Shepard, was a Christian Scientist. Shepard did his primary and secondary schooling from Adams School, Oak Street School and Pinkerton Academy, a private school in East Derry and skipped his sixth and eighth grade. As a boy he was fascinated with flights and he liked to read Charles Lindbergh's autobiographical books. He did odd jobs at the local airfield to learn about airplanes. His father wanted him to join Army but he chose to join Navy. In 1940 at the age of 16, he cleared the entrance exam to the United States Naval Academy at Anapolis but as he was too young to join that year, so the Navy sent him to the Admiral Farragut Academy, which was a private – preschool for Naval Academy. Shepard was an excellent student in academics and outstanding in sports too; he was especially interested in aquatic sports. He was a great and won number of races in sailing. He was so keen on sailing that he learned to sail different types of boats. He was a swimmer as well. Shepard received his Bachelor of Science degree from the United States Naval Academy in 1944, Master of Arts degree from Dartmouth College in 1962, and Honorary Doctorate of Science from Miami University in 1971, and Doctorate of Humanities from Franklin Pierce College in 1972.

He met his future wife Louise Brewer at a party in 1942, when during his Christmas break, he went to Principia College to meet his sister, who was unable to go home due to restrictions to travel as it was wartime. Shepard married Louise on March 03, 1945 in St. Stephen's Lutheran Church in Wilmington, Delaware where his father became his best man at the wedding. He nicknamed her "Saint Louise", for her composure. The couple had two daughters and named them Laura, who was born in 1947, and Julie, born in 1951. They had six grandchildren.

During the closing months of World War II, Shepard served as an ensign on the destroyer Cogswell at the Pacific.

So, eager won shepard to get his wings and his pilot's licence that he studied at a civilian flying school in his spare time while he was attending naval flight training at Corpus Christi, Texas and Pensacola, Florida. Finally he received his Navy wings in 1947, which his proud father pinned on his chest. Next assignment for Shepard was with Fighter

Squadron 42 at Norfolk, Virginia, and Jacksonville, Florida. After World War II Shepard become a test pilot, one who tries out new aircraft and ensures that airplanes are safe. He also worked as an instructor at the Test Pilot School for some time. He logged more than 8,000 hours of flight time in his entire naval career. A great turning point came into his life when in 1959, the newly formed National Aeronautics and Space Administration (NASA) got permission from President Dwight D. Eisenhower to recruit its first astronauts from the ranks of military test pilots. Nearly 508 services records of graduates of test pilot schools were obtained from the Department of Defence. From which 110 were shortlisted based upon the selection criteria of age, height and others, than by going under various level of physical and psychological tests Shepard was also chosen among the seven others, including Scott Carpenter, Gordon Cooper, John Glenn, Wally Schirra, and Deke Slayton for NASA's space flight program. The of group of these seven astronauts came to be known as Mercury 7. The identities of these seven were revealed to public at a press conference at Dolley Madison House in Washington, DC, on April 9, 1959.

After Shepard's flight got rescheduled several times due to various reasons. Shepard finally made history on May 5, 1961, when his Freedom 7 spacecraft flew into the sky from its Florida launch pad, launched by a Redstone vehicle on a ballistic trajectory suborbital flight. The flight took him to an altitude of 116 statute miles and to a landing point 302 statute miles down the Atlantic Missile Range. He became the first American in space, a month after the Soviet cosmonaut Yuri Gagarin had earned the distinction as the first person in space. Shepard's launch was watched live on television by millions of people. He travelled more than 300 miles in his 15-minute-long mission. The mission was a great success and Shepard was treated as a National hero. He came down in the Atlantic Ocean near the Bahamas, where he was picked up by the aircraft carrier Lake Champlain.

Returning to the United States, Shepard came to the White House to receive the NASA Distinguished Service Medal from President John F. Kennedy. He was also honoured with a ticker-tape parade in New York City.

Shepard also served as capsule communicator (CAPCOM) for Glenn's Mercury-Atlas 6 orbital flight and worked on the ground for subsequent flights in the Mercury program and was slated to pilot the Mercury 10 mission. But on June 12, 1963, NASA Administrator James Webb announced that Mercury had accomplished all its goals, and no more missions would be flown. After Mercury 10 mission got cancelled NASA selected Shepard to be the part of the first manned Gemini mission. In early 1964, unfortunately he was diagnosed with Ménière's disease, which is a disorder of the inner ear that can affect hearing and balance. It is characterized by episodes of vertigo, tinnitus, and hearing loss. The hearing loss comes and goes for some time, alternating between ears.

Shepard was designated as Chief of the Astronaut Office in 1963 with responsibility for monitoring the coordination, scheduling, and control of all activities involving NASA astronauts. He helped select new astronauts, plan missions and make sure astronauts were ready to fly. His work also included monitoring the development and implementation of effective training programs to assure the flight readiness of available pilot personnel for assignment to crew positions on manned space flights. He was back to full flight status in May 1969, after a corrective surgery for an inner ear disorder.

In 1971 Shepard got selected for his second spaceflight along with Edgar D. Mitchell after almost 10 years since his first flight. The name of this spaceflight was Apollo 14, which took off on January 31, 1971 which was America's third successful lunar landing mission. This was the first mission to be broadcast successfully over colour television from the surface of the Moon, using a vidicon-tube camera. They spent more than 33 hours on moon, and there also Shepard again made history by playing golf on Moon's surface, since he wanted to show how far the ball can go in the moon's lower gravity. He had packed a specially designed golf club just for this purpose.

Apollo 14 mission was a great achiever in many ways senses. The achievements included: use of first Mobile Equipment Transporter (MET), largest payload placed in lunar orbit, longest distance traversed on the lunar surface, largest payload returned from the lunar surface, longest lunar surface stay time (33 hours), longest lunar surface EVA (9 hours and 17 minutes), first use of shortened lunar orbit rendezvous techniques and first extensive orbital science period conducted during CSM solo operations. He resigned from his duties as Chief of the Astronaut Office in June 1971, and was appointed as a delegate to the 26th United Nations General Assembly by President Richard Nixon in July 1971 serving from September to December 1971. He was promoted to Rear Admiral by Nixon in that particular year. Finally he retired from the Navy and NASA on August 1, 1974. After his retirement Shepard became Chairman of Marathon Construction Corporation and founded his company, Seven Fourteen Enterprises. He also chaired as the President of the Mercury Seven Foundation, which was a non-profit organization which provides college science scholarships for deserving students. He wrote two books with other co-authors, "Moon Shot: The Inside Story of America's Apollo Moon Landings" and "We Seven: By the Astronauts Themselves."

On July 21, 1998, Shepard died at Community Hospital on the Monterey Peninsula because of leukemia (mainly known as blood cancer), in Pebble Beach at the age of 74. He was diagnosed with that disease two years before his death. Within five weeks of his death, his wife also passed away at 76 on August 25. They had been happily married for 53 years and their ashes, were scattered together by a Navy helicopter.

Shepard was associated with many organizations including American Astronautical Society and the Society of Experimental Test Pilots, he was also the member of the Rotary, Kiwanis, the Mayflower Society, the Order of the Cincinnati, and the American Fighter Aces; honorary member. He was also among the Board of Directors for the Houston School for Deaf Children, Director of National Space Institute, and Director, Los Angeles Ear Research Institute. Shepard has a long list of Awards and achievement in his name that include Golden Plate Award for Science and Exploration in 1981, the Langley Gold Medal (highest award of the Smithsonian Institution for Aeronautics and Astronautics) on May 5, 1964; the John J. Montgomery Award in 1963 and many others. In Hampton, Virginia, a road is named Commander Shepard Boulevard in his honour. USNS *Alan Shepard* (T-AKE-3) a Naval ship was named after him in 2006. In Concord, a discovery center is named after Shepard and Christa McAuliffe as "The McAuliffe-Shepard Discovery Center in New Hampshire." Due to Shepard's career as an astronaut, his hometown of Derry has the nickname

Space Town in his honor. There are also many schools named in his memory like Alan B. Shepard Middle School etc. In a space survey in 2010, Alan Shepard was ranked 9th as the most popular space hero. In 2011, NASA also honoured Shepard with an Ambassador of Exploration Award, for his contributions to the U.S. space program. On May 4, 2011, a first-class stamp was also issued in Shepard's honour in U.S. Postal Service. The Space Foundation, in partnership with the Astronauts Memorial Foundation and NASA, present the Alan Shepard Technology in Education Award for outstanding contributions to creative and innovative use of technology in education by K-12 educators or district-level education personnel each year. A book "Light This Candle: The Life and Times of Alan Shepard", which was written by Neal Thompson was published in March 22, 2005, it that covers the entire life of Alan Shepard and his struggles.

Shepard life is a great inspiration to everyone, and teaches us that no matter how hard the life is if we remain focused on our aim, we will definitely taste and his success. He inspired us to be fearless decision. His words : "Whether you are an astronomer or a life scientist, geophysicist, or a pilot, you've got to be there because you believe you are good in your field, and you can contribute, not because you are going to get a lot of fame or whatever when you get back." Shepard's life and his work as an astronaut, and as Naval officer are inspiring and worth remembering.

□

CHRISTA MCAULIFFE
(The Teacher in Space)

Christa McAuliffe was a dynamic and compassionate teacher. She was the first non-astronaut civilian to be selected for a space mission. Christa was sent to space in 1986 for the TEACHER IN SPACE PROJECT. The project was meant to carry teachers in space, and return to their classrooms to share the experience with their students.

Sharon Christa Corrigan McAuliffe was born on September 2, 1948 in Boston, Massachusetts. Her father Edward Christopher Corrigan, was an accountant and her mother, Grace Mary Corrigan, was a substitute teacher, Christa was the eldest of her five siblings.

As soon as Christa was born, her father completed his sophomore year at Boston College and accepted employment as an assistant comptroller with a Boston departmental store and then the family shifted to Framingham, Massachusetts where Christa completed her graduation from Marian High School in 1966. She also met the love of her life Steven J. McAuliffe there. As a young girl, she was very fascinated with Apollo's moon landing program and Project Mercury. She attended Framingham State College and received her Bachelor of Arts degree in 1970. In that same year Christa and Steven were married.

They had to move to Washington D.C. so that Steven

could attend the Georgetown University Law. Christa always had a love for teaching and her first job in 1970 was as an American history teacher at Benjamin Foulois Junior High School in Morningside, Maryland until she delivered her first child, Scott on September 1, 1976. She also taught history and civics at Thomas Johnson Middle School in Lanham, Maryland from 1971-1978. Later in 1978, she attended Bowie State College for a master's degree in school administration and then shifted to Concord where Steven started working as an assistant to the New Hampshire Attorney General. Then Christa gave birth to their second child a daughter named Caroline, on August 24, 1979 and she did not resume teaching but stayed at home to raise Caroline and Scott. In 1983 she resumed teaching at Concord High School. She always believed in practical knowledge than theoretical, and took her students out for field trips so that they could relate themselves with the topic. In her teaching career she taught several courses like American history, law, and economics and she also introduced and taught a course entitled "The American Woman".

Christa was also a social worker and a generous contributor to the community; she liked to participate in all kind of activities from playing tennis to going to church. Her interests included playing tennis, and volleyball, she was a jogger and a swimmer too. Music was an equally important part of her life. She also campaigned for a local hospital and for the YWCA. She joined the Junior Service League and was a Girl Scout troop leader.

Her selection for the space project was not a mere coincidence but the fact that she was interested in space projects since her youth days which she even mentioned in her application form for the space project saying

"There was excitement in my home when the first satellites were launched. I was caught up with their wonder. I cannot join the space program and restart my life as an astronaut, but I watched the Space Age being born and I would like to participate."

When on August 27, 1984 the United States President Ronald Reagan announced the Teacher in Space Project, Christa was teaching at Concord High School. Her interest towards the aircrafts and National Aeronautics and Space Administration's (NASA) space missions motivated her to file her 11 pages application form on 1 February 1985. NASA's goal from this project was to spur student interest in mathematics, science and space exploration.

NASA thought of sending a person to space to kindle interest in the common public. At the time of teacher in space project NASA was under pressure to find financial support so it was also important for it to demonstrate reliability of spaceflight. It was decided that a teacher for this purpose would be the best suited for this purpose as she/he would be the one who could inspire many and would be a better communicator.

Nearby 11,000 application forms were received from interested teachers. A non-profit organization was chosen to coordinate with NASA named as Council of Chief State School Officials. It selected 114 semi-finalists from the initial application round. This selection included two teachers from each state, the District of Columbia, Puerto Rico, the Virgin Islands and the territories and trusts of the United States. The state of New Hampshire nominated two teachers out of whom one was McAuliffe.

The semi-finalists then gathered to attend a conference on space education and to meet the National Preview Panel from June 22-27 in 1985. The semi-finalists then went through various levels of selection. On July 1, 1985, this panel shortlisted 10 teachers among whom stood the first teacher to go in space. The ten finalists then went to NASA's Johnson Space Center in Houston, Texas where they gave their medical fitness test, final interviews and briefings. On July 7 Christa travelled to Johnson Space Centre for medical examinations and gaining information about spaceflights.

The Interview was held for the 10 finalists by an evaluation committee that composed of senior NASA officials who were supposed to make recommendations to the NASA administrator James M. Beggs for the selection of primary and backup candidates for the project.

Christa's selection to represent the first teacher in space was announced on July 19, 1985 by the Vice President George H.W. Bush at a White House ceremony. Barbra Morgan a third-grade teacher in McCall, Idaho, was selected as her backup for the, "Teacher in Space Project". On July 19, 1985 Christa also received a keepsake award at a formal announcement ceremony at the Johnson Space Center in Houston, Texas.

NASA official Alan Ladwig was quoted as saying, "She had an infectious enthusiasm" and NASA psychiatrist told a famous magazine that "She was the most broad-based, best-balanced person of the 10 finalists."

After the formal announcement of Christa's selection on July 1985, she returned to Concord on August 6, 1985, where the city celebrated 'Christa McAuliffe Day' in her honour. She also received formal appreciation and praise from her town and from the state of New Hampshire. The National Education Association (NEA) also honored Christa for her extraordinary achievement.

Both Christa and Morgan had to take a break from teaching for a year in order to train for the spaceshuttle mission in early 1986. They began their astronaut intensive training at the Johnson Space Center for NASA Mission 51-L. In the course of her training she learnt to read flight data and to operate certain cockpit controls. She and Barbra Morgan also practiced for proper entering and exiting the spaceshuttle and learned to operate the ship's onboard cameras. She was trained on a KC-135 training jet that pretended weightlessness for the astronauts. She was trained in such simulators that depicted the appearance of space and the feeling of extra gravity pull on lift-off. Christa emergency training included firefighting and the use of a ,"Rescue ball (which is a solution for transporting astronauts from one Space Shuttle to another in case of an emergency)". Christa learned to operate galley equipment, and even how to accomplish day to day activities and bathroom operations in outer space. Her flight clothing included shirts, shorts, underwear, socks, flight boots, gloves, pants, a jacket, coveralls and a personal hygiene kit. She also carried a supply kit containing watch, flashlight, pressurized pens, pencils, sunglasses, scissors, a pocketknife, earplugs etc. She learned to operate a sleep restraint harness to prevent drifting about the cabin when resting. She also brought six cassettes and a tape player for entertainment. Christa was also trained to capture clear, sharp, detailed photographs from space with a personal camera.

The training lasted for 120 days and throughout the astronaut training with the help of media, McAuliffe shared her experience with American people in forms of Media Outlets.

Christa prepared two in-flight nothings outer to teach

lessons from space which included experiments in the fields of chromatography, hydroponics, magnetism and Newton's laws. She had also planned to conduct a tour of the spacecraft called 'The Ultimate Field Trip' and to teach the benefits of travelling to space called "Where We has Been, Where We're Going, Why." These lessons were to be broadcast for millions of school children. Besides the lessons Christa also maintained a personal journal in order to record her thoughts where she could write about all her experiences. "That's our new frontier out there, and it's everybody's business to know about space."

She had always been a source of inspiration for her students and fellow teachers. She was invited as a guest on several T.V. shows like Today Show, CBS Morning News, and Good Morning America etc. She once said on record "If you are being offered a seat on a rocket ship, don't ask what seat. Just get on", She immediately gained popularity in media as the first teacher in space which further rekindled the interest in US citizens for the launching projects. This fulfilled NASA's purpose of sending a teacher in space.

On January 28, 1986 Challenger was finally ready for its launch after several rescheduling due to weather and other problems. Christa, along with the other six crew members including Gregory Jarvis, Judith Resnik, Francis Scobee, Ronald McNair, Mike Smith and Ellison Onizuka, boarded the Challenger. It was launched from Pad 39B at the Kennedy Space Center. At an altitude of 48,000 feet at 11:39 am after 73 seconds of its flight the shuttle broke apart, ***burning nearly 2 million liters of fuel in just a few seconds, creating a sinister cloud of gas.*** **All seven crew members died on the spot in front *of their family, friends, and millions of TV viewers,***

who were watching the launch live on their TV Sets.

A commission known as Rogers Commission was set up by the presidential commission to enquire the reasons for the crash of the spaceshuttle. It was found that the reason behind the crash was a flaw in the designing of the shuttle. The Challenger explosion was later determined to be the failure of an O-ring due to extreme cold temperatures.

After her death Christa was buried in her hometown, Concord at Blossom Hill Cemetery. Since then she has been honoured at many events. Many schools and colleges all-around the world have been named after her name also several scholarships are being offered by her. Since 1986 every year, The Christa McAuliffe Technology Conference has been held in Nashua, New Hampshire, and is devoted to the use of technology in all aspects of education.

Later, in 1998, "The Teacher in Space Project", was replaced by Educator Astronaut Project; the first Educator Astronaut was Christa's backup for the Challenger flight, Barbara Morgan, who was launched aboard the Shuttle Endeavour on August 8, 2007.

Christa's husband Steve remarried and became a federal judge at Concord. Her son, Steve, graduated in marine biology and her daughter, Caroline, followed her mother's footsteps and chose the profession of a teacher.

Christa has been a source of inspiration to her students and all others who believe in dreaming and try their best to make them come true. She always believed in living her life with joy and happiness. Her love for teaching always motivated her to try new things. Christa had a simple motto in life, "Reach for it. Push yourself as far as you can."

On July 23, 2004, Christa along with the other astronauts who lost their lives in the challenger were honoured with the Congressional Space Medal of Honour by President George W. Bush.

Christa's mother, Grace Corrigan, in her book, "A Journal for Christa" wrote that "Christa lived. She never just sat back and existed." Christa McAuliffe was an exceptional lady who displayed great enthusiasm and kept the courage of achieving whatever she wished for. She was a gifted teacher.

□

JIM LOVELL
(The Astronaut Who Couldn't Land on Moon)

Jim Lovell is a retired American astronaut of NASA. He is associated with many historical flights such as Gemini 7, Gemini 12 and Apollo 8. He is much remembered for the flight of Apollo 13 that could not land on moon due to some fault but was successfully rescued. He is among the only three people to fly to the Moon twice. He is also the only person who has been to moon twice but without landing on it. For years, he held the record of world's most-travelled astronaut and he has been to space for four times. Before joining NASA, he was in Navy and after his retirement both from Navy and space program, he started working in private sector as a businessman.

James Arthur 'Jim' Lovell, Jr. was born on March 25, 1928 at Cleveland, Ohio, US. His father's name was James Lovell Sr., who died in a car accident when Lovell was only five years old. His mother Blanche, raised her as a single mother Milwaukee, Wisconsin, US. She was a native to the Czech Republic in Central Europe, and Jim was her only child.

Lovell joined Juneau High School at Wisconsin and became an Eagle Scout. From his childhood Jim was fascinated about rockets and flying models. He was a bright and intelligent student and loved to know about astronomy, physics and engineering.

He also attended the University of Wisconsin–Madison for two years under the 'Flying Midshipman', program. He got his Bachelor degree in 1952 from University of Wisconsin; United States Naval Academy, after his graduation from Naval Academy, he underwent flight training from 1952 to 1954 at NAS Pensacola. On June 6, 1952 he got married to Marilyn Lillie Gerlach, who was his high-school friend from Juneau High School in Milwaukee. They dated for a long period of time, earlier Marilyn was hesitant about dating Lovell as he was two years older than her but later they became so much involved that to live near him training in Annapolis she took transfer from Wisconsin State Teachers College to George Washington University in Washington D.C. They have four children: Barbara L (born 13 Oct 1953), James A, Jr. (born 15 Feb 1955), Susan K (born 14 July 1958), Jeffrey (born 14 Jan 1966), and have ten grandchildren. He further continued and completed Advanced Management Program, Harvard Business School in 1978.

After completion of pilot training Lovell served at sea flying F2H Banshee night fighters till 1957. He then entered a six-month test pilot training course in January 1958, at Naval Air Test Center (which is Naval Test Piolet School now). In 1958 Lovell was amongst three of 110 military test pilots selected as potential astronaut candidates for Project Mercury by NASA along with Conrad and Schirra. Out of the three Schirra managed to become one of the member of Mercury Seven Mission, but Lovell and Conrad failed due to the medical reasons. Then four years after Lovell continued to serve as a test pilot instructor and safety officer at Pax River he was called with a nickname given to him by Conrad 'Shaky'.

In 1962, it was announced that NASA had once again begun looking for a second group of astronauts for the

Gemini and Apollo programs. Being rejected earlier Lovell did not lose hope and applied for the second time and was selected for the second group of astronaut along with Conrad. At first Lovell was selected as a backup pilot for Gemini 4 project and finally as a prime member for Gemini 7 with Frank Borman as commander which was launched on 4 December 1965. This project was marked as the longest time spent by humans in space until 1970 with almost 14 days near about 330 hours and 35 minutes in space. After 11 days of its launch Gemini 7 was also joined in space by Gemini 6 which was manned by Walter M. Schirra, Jr., and Thomas P.

Later Lovell was picked to be the backup for Command Pilot of Gemini 10, but after the deaths of prime crew of the Gemini 9 Elliot See and Charles Bassett, he replaced Thomas P. Stafford and became backup commander of Gemini 9A.

His performance on Gemini 7 bought Lovell a commanding position on Gemini 12 (November 11-15, 1966) with Edwin 'Buzz' Aldrin as the pilot. The mission took 4-days, 59-revolutions of the earth in 4 days and was a success including spacewalk by Aldrin. Gemini 12 was successful closure of the NASA's Gemini mission. Now NASA started preparing for Apollo program and the journey to the moon. Lovell was selected as a backup for Neil Armstrong for the Apollo 11 lunar landing mission.

Lovell also served as the Command Module Pilot and the Navigator, using the spacecraft's built-in sextant to determine its position by measuring star positions on the epic six-day journey of Apollo 8 – man's maiden voyage to the moon – December 21-27, 1968. Apollo 8 was the first manned spacecraft to be lifted into near-earth orbit; and Lovell and fellow crew members, Frank Borman and

William A. Anders became the first humans to leave the Earth's gravitational influence. On Christmas Eve, the Apollo 8 astronauts described the moon and then read from the book of Genesis during a live television broadcast from space. Later the reading brought on a lawsuit against NASA.

Lovell was scheduled to command Apollo 14, but the crew of Apollo 13 was swapped with Apollo 14, as the commander of Apollo 13 Alan Shepard, needed some more time to get trained since he has been grounded for a long period of time due to medical reasons.

Lovell completed his fourth mission as Spacecraft Commander of the Apollo 13 flight, April 11-17, 1970, with CM Pilot Jack Swigert and LM Pilot Fred Haise. It was decided that he and Haise will land on the Moon. With this launch Lovell became the first man to journey twice to the moon. Apollo 13 was planned for ten days but on April 13, two days after its launch during a routine cryogenic oxygen tank stir of Apollo 13 on its way to moon, suddenly the damaged electrical insulation on the wiring created a spark and the tank exploded, causing a loss of oxygen and started a fire inside the tank.

Liquid oxygen rapidly turned into a high-pressure gas, which burst the tank and caused the leak of a second oxygen tank. These explosions and leaks led to shortage of power and oxygen and forced the abandonment of the Moon mission. The only motive of NASA's scientist on the earth was now to rescue the crew members of Apollo 13 safely. The rescue mission involved the use of the Lunar Model as a 'life boat', providing battery power, oxygen, and propulsion, which helped Lovell and other crew members to land safely on the earth. The mission was a success and by the hard efforts of NASA accompanied by the prayers of the million people Apollo returned on the earth safely on April 17, 1970. The mission was known as a 'successful failure', because the crew was brought back to earth safely. The Apollo 13 tragedy also made Lovell, Haise, and Swigert the men who had travelled the farthest distance from Earth.

Along with John Young and Eugene Cernan, Lovell is one of only three men to travel to the Moon twice, but unlike them was never able to walk on it. Lovell also held the record for spending maximum time in space for a long period with a total of 715 hours and 5 minutes until it was surpassed by the Skylab flights in 1973. Jim Lovell has witnessed 269 sunrises from space on his Gemini and Apollo flights.

Lovell took his retirement from Navy and the space program in 1973 and started working for Bay-Houston Towing Company in Houston, Texas, and became its CEO in 1975.

On January 1, 1977, Lovell became President of Fisk Telephone Systems (marketing business communications equipment), in the southwestern United States. On January 1, 1981, he was appointed Group Vice President, Business Communications Systems, later worked for Centel, retiring as an executive vice president on January 1, 1991.

Lovell also authored a book along with Jeffrey Kluger, about the Apollo 13 mission, Lost Moon: The Perilous Voyage of Apollo 13. Based on the book the Apollo 13 mission was portrayed in the 1995 film, Apollo 13. Tom Hanks played the role of Lovell and her wife's role was played by actress Kathleen Quinlan who got nominated for the Academy Award for Best Supporting Actress for her performance as Marilyn Lovell. He also played a cameo as the captain of the USS Iwo Jima. His wife too has a cameo in the film. He was portrayed as one of the astronauts featured in the book and documentaries 'In the Shadow of the Moon' and 'When We Left Earth'. In 1999, the Lovell family opened a restaurant that displayed many artifacts from his time with NASA, as well as from the filming of Apollo 13. The name of the restaurant was 'Lovell's of Lake Forest', in Lake Forest, Illinois.

After his retirement Lovell visited many college, schools and universities, where he shared his experience and learning as an astronaut and motivated students and youth to get involved in science and the space program. Now at the age of 87 also, Lovell still travels for space events and sometimes gives interviews.

There are many awards and honour, which Lovell has recorded in his name. A small crater on the far side of the Moon is named in his honour, many streets are also named after him including North James Lovell Street, the stretch of 7th Street between W. State Street and W. Clybourn Street in downtown Milwaukee, Wisconsin. Many schools and colleges are named after him to honour him. The Captain James A. Lovell Federal Health Care Center is also named in his honour. In Milwaukee, Discovery World was named, "The James Lovell Museum of Science, Economics and Technology," which too is located on James Lovell Street, named after him. He has received many awards too including the Robert J. Collier Trophy, Harmon International Trophy (three times) and Goddard trophies, the Presidential Medal of Freedom, NASA Distinguished Service Medal and the Congressional Space Medal of Honour. He has received Distinguished Eagle Scout Award in 1976, Silver Buffalo (Boy Scouts of America) in 1992. In Space Foundation survey in 2010, Lovell was ranked amongst the four most popular space heroes. He has also been named the person of the year by *Time* Magazine in 1968 along with Apollo

8 Astronauts Frank Borman and William Anders. He has been on the cover page of the Time magazine twice (1969 and 1970) and on the cover page of *Live* Magazine once on 24 April 1970. In 2013 he received, "The Honourable Company of Air Pilots Award of Honour", presented by His Royal Highness the Duke of York along with other numerous awards received by him.

Lovell is a very hard working and consisted person and he never believed in giving up. He inspires us to keep working chance. He proved that failure is also a step toward success, no one should be disheartened and should never lose hope, instead should learn from failure and should gain lessons from rejections and work hard in order to deal with them.

He once said, "Be thankful for problems. If they were less difficult, someone with less ability might have your job."

Lovell has shown great qualities of leadership and humanity. During the Apollo 13 mishap he acted very calmly, patiently and wisely.

"There are people who make things happen, there are people who watch things happen, and there are people who wonder what happened. To be successful, you need to be a person who makes things happen," he was quoted.

□

MARC GARNEAU

(The First Canadian in Space)

Marc Garneau, the Canadian astronaut is person multi talents. During his career, he served as an astronaut, military officer, engineer, instructor, President of the Canadian Space Agency and as a politician too; he did his best in every field. On October 1984 he became the first Canadian Astronaut to fly in space as a payload specialist on Shuttle Mission 41-G, October, and also became the Canadian who has been to space thrice. He is a person with positive approach, is brave and smart in making decisions. He pursued all his duties with honesty and responsibility.

Marc Garneau, now 67 years old was born on February 23, 1949 at Quebec City, Quebec, Canada. His father Brigadier General André Garneau, was a career infantry soldier who fought during the Second World War and his mother name was Jean Garneau who was an Anglophone from New Brunswick. He had one elder brother and two younger brothers. His full name is Joseph Jean-Pierre Marc Garneau. He completed his early schooling from schools in Quebec City and Saint-Jean-sur-Richelieu and earned his Bachelor of Science degree in Engineering Physics from the Royal Military College of Kingston in 1970. In 1973 he did his Doctorate in Electrical Engineering from the Imperial College of Science and Technology, London, England.

He had a love for science from his early age. He liked to explore new things and gain knowledge about the things in his environment. As a boy he was attracted to sea and that was the reason he joined Navy later on. He also attended the Canadian Forces Command and Staff College of Toronto in 1982-1983. From 1974-76 he worked as a Navy Combat Systems Engineer in Royal Canadian Navy on HMCS Algonquin. While he was working as an instructor in the naval weapon systems at the Canadian Forces Fleet School at Halifax, he also designed a simulator which could be used in training weapons officers for using missile systems aboard Tribal class destroyers. He also played a significant role in the development of an aircraft-towed target system for scoring the accuracy of naval gunnery exercises. He also served as a project engineer from 1977 to 1980 at Ottawa in naval weapon systems. In 1982 he was promoted as a commander due to his dynamic personality and enormous knowledge, he soon became a well-known authority in the design of electronic warfare equipment and systems. While working in the forces, he helped the Canadian Forces to develop numerous items that assisted with communications, warfare equipment, and superior naval weapons. He got promoted to the rank of captain in 1986 and finally retired from Navy in 1989.

Since childhood Marc had been a focused person with clear goals and plans in his life. He enjoys flying, scuba diving, squash, tennis, car mechanics, cycling and walking around to remain fit. On a 59-foot yawl with twelve other crewmen he also has sailed across the Atlantic twice, in 1969 and 1970. He married his second wife, Pamela Soame from Ottawa, in 1992; after his first wife Jacqueline Brown, from whom he had separated, committed suicide in 1987. Marc and Pamela have two children; Adrien and George. Garneau also had twins from his first marriage; Yves and Simone. He and Pamela have been residents of Westmount since 2001.

One of his dream came true when in 1983, "The Canadian Astronaut Program", was established under, the management of the National Research Council of Canada. They called for the applications from those interested for their first mission and becoming the Canadian astronauts.

In response to the first call for Canadian astronauts, more than 4,000 applications were received by the Canadian Astronaut Program. After an exhaustive evaluation period, Marc Garneau, Roberta Bondar, Steve MacLean, Ken Money, Bob Thirsk, and Bjarni Tryggvason were selected based on their excellent academic backgrounds, professional experience, health, communication skills and other necessary criteria. Garneau became the first Canadian Astronauts and also first Canadian in outer space. From October 5 to 13 in 1984, he flew on the shuttle Challenger, STS-41-G, which was the 13th flight of NASA's Space Shuttle program and the sixth flight of Space Shuttle Challenger. It was launched on October 5, 1984 from the Kennedy Space Center at 7:03 am EDT, mission STS-41-G became the historical shuttle mission in many sense, it was the first shuttle mission to carry a crew of seven members, including the first crew with two women with it (Sally Ride and Kathryn Sullivan), the first American Extravehicular activity involving a woman (Sullivan), the shuttle was also carrying a first Australian-born person to journey into space and the first astronaut with a beard (Paul Scully-Power), and the first Canadian astronaut (Marc Garneau). Marc was designated as the payload specialist of the STS-41G mission of the American space shuttle Challenger. Astronaut Robert

Thirsk was trained as his backup for the mission, the main objective of this mission included the demonstration of the Orbital Refuelling System by Sullivan and Leestma during a spacewalk and the deployment of the Earth Radiation Budget Satellite.

One hundred thirty-two orbits of earth were completed by STS-41G in 197.5 hours, before its landing at Kennedy Space Center, Florida, on October 13. In space Marc Garneau conducted different types of experiments under three main categories: space technology, space science and life sciences. He also first time tested the space vision system designed to provide eyes to the robotic arm of the shuttle.

A law was adopted by the Canadian Government creating Canadian Space Agency (CSA) to coordinate all the country's space activities. To devote himself entirely to his career as an astronaut, Marc left army. He became deputy director of the Canadian Astronaut Program from 1989 to 1992. In July 1992 with the US National Aeronautics and Space Administration (NASA), he was selected for astronaut training and reported to the Johnson Space Center in Houston, Texas, for mission specialist training and after undergoing a year of rigorous training, he was qualified to command the orbiter.

He was also assigned to CAPCOM (Capsule Communicator) duties at NASA for a number of shuttle flight, thus becoming the first non-American to communicate with the shuttle crew from Mission Control. Garneau logged more than 677 hours in space, he boarded three space flights—STS-41G aboard Challenger in 1984, STS-77 on Endeavour in 1996 and STS-97, also onboard Endeavour, in 2000.

Garneau was appointed as an executive vice president of the Canadian Space Agency in February 2001, and in

November 22 of the same year, he became president of CSA. On 28 November 2005, he resigned from his position, to run for office in the federal election. He worked as an astronaut almost for 15 years, making countless contributions to the Canadian space program.

His political career started in 2006, when he became a Liberal candidate for Vaudreuil-Soulanges, which is a constituency in western Québec, but unfortunately he was defeated by incumbent Meili Faille of the Bloc Québécois. Winning by 9,000 votes in 2008, Garneau was elected in the Montréal riding of Westmount–Ville-Marie. Despite of his non-experience in politics, but only due to his hard work, will power and intelligence, he became among one of the leading members of Michael Ignatieff's Québec caucus. In 2011 Garneau was narrowly re-elected in 2011 after he beat New Democratic Party candidate Joanne Corbeil. He then served as the Liberal house leader, responsible for his party's strategies and tactics.

On November 21, 2012, Garneau was named as his party's Natural Resources critic after David McGuinty resigned the post but on January 30, 2013 he was replaced by Ted Hsu.

In the election held on October 19, 2015, Garneau was re-elected in the newly created new federal electoral district in Quebec, on November 4, 2015, Garneau was appointed as a Transport Minister of Canada in the federal Cabinet headed by Justin Trudeau. He is a pegler leader and frequently meets with individual citizens and associations to know and take account about their concerns and to answer their questions so that he can more effectively represent the people and can resolve their issues and needs. He is currently serving as Liberal critic for Foreign Affairs, International

Trade and La Francophonie.

Garneau has won a long list of awards and honours in his name. To recognize his role as the first Canadian astronaut, he was appointed as an Officer of the Order of Canada in 1984, Canada's highest civilian award. He was awarded with Exceptional Service Medal by NASA in 1997, He was promoted to the rank of Companion in 2003 for his remarkable work with Canada's space program. He received Golden Jubilee Medal in 2002. He was also awarded for his 12 years of honourable service as the Canadian Forces Decoration with the Canadian Forces. A high school is named Marc Garneau Collegiate Institute in Toronto, in the honour of Marc Garneau and a school is named École Secondaire Marc Garneau in Trenton, Ontario in his honour. On 10 December 1984 he was awarded the Key to the City of Ottawa from Marion Dewar the Mayor of Ottawa, Baldwin Award in 1985 for the best paper in the Canadian Aeronautics and Space Journal and so on.

He also continued his studies and took several degrees Doctor of Law, Doctor of Science, Doctor of Technology etc. in 2006.

Garneau is an inspiration to everyone around the world in order to touch the success, and he shows us that, Having a positive attitude and willingness to do hard work are the only secrets to success. He has been a passionate educator and inspirer for space exploration. A spaceship was named USS Garneau after him in the popular Star Trek series of novels, this shows how profound influence Marc has had on his fans of space discovery.

Once he said that, "I have to be a good role model for the rest of my life. That's a lot of pressure because if I screw up once, I'm going to disillusion a lot of people."

Marc dedicated his entire professional life to his country by serving Canadian Navy and then as astronaut and President of the Canadian Space Agency, and continues to do serve his land as an active politician.

As a dedicated politician Marc is hard at work for the welfare of the people he represents. He readily mingles with associations and meets up individual citizens to have a first-hand account of their concerns. Since his first winning in 2008, he had always strived to serve the residents of Westmount-Ville-Marie and give them an effective representation.

□

VALERI POLYAKOV

(The Astronaut Who Spent Maximum Number of Days in Space in a Single Stretch)

Valeri Vladimir Polyakov is former Russian cosmonaut. He is known for holding the record of living aboard space station for the longest stretch in single space flight. Vladimir has lived in the Soviet space station Mir for above a period of 14 months or precisely for 437 days and 18 hours during one space trips. The total time spent by him in space during various flights is of 22 months.

He became a cosmonaut in 1972 and his maiden flight to space came in 1988 when went aboard Soyuz TM-6, under the mission Mir EO-3/Mir EO-4, only to return to the earth after 240 days aboard TM-7. His second mission to space was Mir EO-15/Mir EO-16 and the duration of his second flight to space spanned across 1994-1995. He spent 437 days in the Soviet space station Mir during this mission. His launch vehicle for this long journey to space was Soyuz TM-18 and return spacecraft was TM-20.

Valeri Vladimirovich Polyakov was born on April 27, 1942 and named Valeri Ivanovich Korshunov by his parents. Polyakov was born and brought up in Tula, Tula Oblast in Russian Soviet Federative Socialist Republic. His stepfather adopted him in 1957 and his name was legally changed Valeri Ivanovich Korshunov to his present name, Valeri

Vladimirovich Polyakov. Polyakov went to the Tula Secondary School No. 4 for his early education and graduated from there in 1959. After school he enrolled for a doctoral degree in the I. M. Sechenov 1st Moscow Medical Institute. Later he joined the Institute of Medical and Biological Problems, the leading Soviet institution for space biomedicine under the Ministry of Public Health, Moscow. Here his specialization field was astronautics medicine. In 1972 he passed his exams to become one of the first doctor-cosmonaut trainees from the institute. He earned a Candidate of Medical Sciences degree in 1976.

Polyakov was fascinated by the news of the first physician Boris Yegorov flying to space aboard Voskhod. The news made him decide to pursue his interest of space medicine.

His interest in astronautics medicine segued to become a cosmonaut. Dr. Valeri Polyakov joined the Medical group 3 of Russia's space program on March 22, 1972, in the capacity of a cosmonaut. He became part of the Russian cosmonaut team as a practicing physician and was trained to provide all kinds of medical assistance to the cosmonauts in orbit, including the surgical assistance if required. A specific spaceflight medical training was provided to Dr. Valeri Polyakov to enable him to offer medical support work for the crews of Soyuz spacecraft and the Salyut space station.

The final review list of Voskhod flight candidates finalized on 29 May 1964 included Polyakov's name along with other four cosmonauts. The General Staff ordered Lazarev, Yegorov, Polyakov, Sorokin, and Katys to enter training for flight aboard the Voskhod spacecraft. During the training Benderov hemorrhaged excessively during centrifuge training and had to withdraw. Polyakov also lost his chance to go aboard the Voskhod spacecraft as he reacted very poorly to the barometric chamber.

For more than a decade and half Ployakov served as reserve cosmonaut for various crew members on different space missions. His first chance to go to space came in 1988 as the research-cosmonaut. He was launched on August 29, 1988 onboard Soyuz TM-6. The spacecraft docked with the Mir space station and Polyakov made the space station his home for 241 days. His schedule during his long stay in space included conducting numerous medical experiments and studying the effects of microgravity on humans. He made his journey back after 240 days 22 hours and 34 minutes and landed on the earth on April 29, in Soyuz TM-7.

Back on the earth Polyakov continued to be an important part of Russia's space programs and served as the Medical Deputy of the Flight Director and also headed a project of IBMP (Institute of Bio Medical Problems). The project was commissioned to refine the strategy of the executive medical support of future missions to the Mir space station.

Russia's space programs had always focused upon studying the effects of microgravity conditions on the cosmonauts during and after their long stay in the orbit. Russian cosmonauts had typically spent more time aloft in space than their American astronaut counterparts. Polyakov volunteered for the next long duration space mission to learn the response of the human body during a long duration mission to Mars under the microgravity environment. In his second spaceflight he flew as the doctor – cosmonaut of Soyuz TM-18 to stay on the Mir space station from January 8, 1994, to March 22, 1995 spending 438 days in space in single flight. This mission made Polyakov a record holder of the longest continuous stay in space, which stands until today. During this journey he orbited the earth 7,075 times covering a distance of 186,887,000 miles. Through this space odyssey Polyakov also broke the space flight duration record set six years by Vladimir Titov and Musa Manarov.

The record for cumulative days spent in all space flights, however, was later broken by Sergei Avdeyev, who spent 188 days, 179 days, and 379 days in three separate space flights between 1992 to 1999. The record is currently held by Gennady Padalka who has spent 879 days spread across five space flights, while living in the Russian Space Station Mir and International Space Station. He is an active cosmonaut unlike Polyakov and Avdeyev who are retired now.

Polyakov's historical mission did not have an auspicious start. As the spacecraft that dropped him onboard Mir, it grazed the space station while the cosmonauts onboard the craft attempted to take pictures of Mir. Fortunately no threatening damage was done to the space station which would be home to Polyakov for more than a year.

Rest of Polyakov's stay onboard Mir was relatively uneventful minus the initial hiccups of adjusting in a long duration space flight. After first three weeks, his mental performance was recorded to be the same as should be on earth.

His schedule onboard Mir included conducting medical, physiological and sanitary-hygienic researches. Some of the researches and experiments were also the components of international space medicine projects. Data collected during Polyakov's flight was later used by the researchers on ground to conclude that humans are capable of maintaining a healthy mental state during long-duration space flights.

NASA astronaut Norman Thagard, who met Polyakov onboard Mir when he was about to leave for the earth after long stay in the space station, observed that Polyakov looked "big and strong" and "like he could wrestle a bear," and that this became possible due to the regular strenuous workout regimen followed in space.

Polyakov took great pains to show the journalists, media persons and onlookers that he suffered no bad effects of zero gravity as he came out of the capsule after landing on the earth. Polyakov denied the helpers who offered to carry him for a few feet from the Soyuz capsule to a nearby lawn chair, but he walked the same distance. In doing so, he wished to prove that humans could be physically capable of working on the surface of Mars after a long-duration transit phase.

Philip Baker, in his book, 'The Story of Manned Space Stations', wrote "When his capsule landed in Kazakhstan he walked from it to a nearby chair, a tremendous achievement...." "He also stole a cigarette from a friend nearby, but could hardly be blamed for that. He sipped a small brandy and inwardly celebrated his mission. His record still stands, and it is unlikely to be broken until man ventures to Mars." Reportedly, his first statement back on Earth was to tell a fellow cosmonaut, "We can fly to Mars."

In an interview given to *New York Times*, in 2009 he insisted that his return from the space after long stay in the space station was in safe and healthy conditions and shows that, "It is possible to preserve your physical and psychological health throughout a mission similar in length to a flight to Mars and back."

Polyakov was under continuous medical supervision before, during and after the flight, with assessments recorded at each step. Comparisons of the collected assessments from the pre-flight, in-flight, post-flight and the two follow-up assessments 6 months after the mission revealed that in Polyakov's case cognitive or mental functions, such as thinking, feeling, sensation etc. were not impaired. However, Polyakov experienced unhappy moods, feelings of raised workload and disturbed patterns of performance tracking and time-sharing during first three weeks in space and in first two week after returning to the earth. Also it was noted Polyakov's mood and performance remarkably stabilized and were same as pre-flight baseline levels during the second to fourteenth month of residence in space. Researchers after the follow up performance assessment on the earth also concluded that there were no long lasting deficits in performance.

An important conclusion drawn after Polyakov's long duration space stay was that it was possible to maintain mental efficiency and stable mood even during extraordinary long-term space missions. However, first three weeks during long duration space flight and first two weeks on the earth after return to earth are crucial when adverse effects may be expected due to the extreme environmental changes. The findings were especially useful for planning manned missions to other planets including Mars.

Following his long duration space flights Dr. Polyakov was actively involved in the discipline of international space flight and became a, "cosmonaut-investigator", for the United States, Austria, Germany, and France during their respective space science missions to the Mir space station. His experience as a cosmonaut as well as in the field of astromedicine has helped in fostering international cooperation along with valuable contribution towards making international space programs medically safe.

Polyakov retired from the Russian space service on

June 1, 1995 and quit his active position of a cosmonaut. He retains his duties as deputy director of the Institute for Biomedical Problems, a post to which he had been appointed in 1989. Later in 1999 he participated in the "Simulation of Flight of International Crew on Space Station" (experiment SFINCSS-99). He is married and has one child. Although, Polyakov is not an active cosmonaut now but he is deputy chairman of the Russian Chief Medical Commission, and participates in the qualification and selection of cosmonauts. He also holds membership in the International Space Researcher's Association and the International Academy of Astronautics. Currently Polyakov holds the position of the Deputy Director of the Ministry of Public Health in Moscow. He and his team are dedicated to the cause of studying the medical aspects of long duration space missions.

Polyakov is decorated with several honours and awards for academic achievements and record making long duration space flights. Some of his awards are the Hero of the Soviet Union/Russian Federation, Order of Lenin, Order of the Legion of Honour, and the Highest Kazakhstan Award, Order of Parasat. He also holds the title of , "Pilot-Cosmonaut of the USSR. Polyakov has over 50 published works to his name. His writings pertains to the field of medical aspects of space missions, life sciences and the research results of long duration space flights."

Despite Polyakov making history and volunteering for the experiment involving lost duration space stay rumours are rife that Roskosmos never made public Polyakov's around 50 medical publications, which were based on his experiments in space.

In most space documentaries, whenever he is filmed viewers continue to see him in sitting position always. Film maker Dana Ranga referred to him as "fragile" and in his congratulatory video to the Mars-500 crew he can never been seen standing or moving around.

It is often speculated that in a few instances if he is ever photographed standing, he may have been propped up. It is also an observation that Polyakov did not walk to the podium at the state event of museum opening just like other fellow cosmonauts who had spent lesser time at a stretch in space.

The data collected from the Mars-500 suggests that long-term space journeys can prove actually detrimental to the space passengers' health and mental conditions. If findings are to be believed it is proved that even the carefully selected, high-performing motivated crews when put into isolation may show symptoms of sleep disturbances, episodic depression, and culturally-driven behavioural challenges. Also it is a proven fact that astronauts spending more time in space lose their bone density faster; even faster than the women on the earth who suffer from osteoporosis. Although bone density can be restored to great extent successfully under medical guidance, it is difficult to regain the same strength in the hip joints.

Mystery continues to shroud this record making and record breaking cosmonaut, mainly due to the fact that he is rarely seen in public.

□

ALEXEY LEONOV

[The First Man to take a Spacewalk for 12 min (the Astronaut Who Shot Video while Walking in Space)

Alexey Leonov made history at the age of 31 when he became the first person on earth to walk in space. He was born on 30th May 1934, in Listvyanka in Central Siberia.

An artist at heart, Leonov completed his graduation with honours from the Chuguyev Higher Air Force School in 1957 and was commissioned as a fighter pilot. Later he attended Zhukovskiy Air Force Engineering Academy in 1968. Soon after he completed his graduation in his early twenties, he was invited in 1959 to try his chances to go to space and was chosen along with Yuri Gagarin, to be part of the first 20 cosmonauts. What followed after the selection was months of hard work and rigorous training. In the early 1963, Leonov was picked up and selected from the group by Sergei Korolev, the mastermind behind the Russia's space program to lead the special mission called Voskhod (Sunrise) 2 that was designed with the sole objective of Spacewalk.

Some four decades later Leonov shared the interesting story behind his selection for the special program, he said:

"So, he (Sergei Korolev) looked over our team of 20, and he stared at me and said: "Hey, ace, you put on the space suit." He told me to try doing the full routine – spaceship, airlock, outside, airlock. "Also, submit your report to the technical command in two hours." Of course, this was one of the most important tasks in my life. Later, I often thought whether it had been a mere coincidence or not. Why did he choose me out of these 20 guys? Gagarin was there, too. In fact, it wasn't by accident. Korolev said a lot of good things about me at the command meeting. He had studied my CV carefully. But I only learned about it 30 years later, when I read the minutes of that meeting. This is how I made it to this program."

The selection followed an extensive training of 18 months on weightlessness. He was accompanied by his friend and colleague Pavel Belyayev on Voskhod 2. The duo learnt to fly and navigate the spacecraft before the historic day. Fifty-one years ago, on 18th March 1965, Leonov stepped outside his Voskhod 2 spacecraft, connected to the spacecraft by a 16 feet long tether and successfully accomplished the mission by walking in space for 12 minutes and 9 seconds. He later described his feelings in the words, "My feeling was that I was a grain of sand." Before he could come back into the spacecraft, Leonov had a tough time in the space when the spacesuit he was wearing got inflated and began to balloon out of shape. The shared his experience with the Observer, "I was surrounded by stars and was floating without much control. I will never forget the moment. I also felt an incredible sense of responsibility. Of course, I did not know that I was about to experience the most difficult moments of my life – getting back into the capsule." After his spacesuit got inflated, his feet came out of the boots, his hands slipped out of his gloves and he could not re-enter his spacecraft's airlock. However, Leonov somehow managed to bleed some air out of his suit by opening a valve which gave him the opportunity to get back into his spacecraft. That was not the end of worries on the mission for Leonov and Belyayev. As soon as the cosmonauts fired small explosive charges to eject the airlock into outer space, the spacecraft started to rotate and before they could actually do something to control that, the oxygen pressure inside the spacecraft started shooting up to dangerously high levels. Any spark inside the spacecraft at that moment of time would have proved disastrous. Once again, when everything appeared to settle down with the oxygen levels receding and they were preparing to return on earth, the automatic re-entry mechanism of the spacecraft failed to work forcing the cosmonauts to operate the system manually. Surviving all odds, the Voskhod 2 landed safely in the deep and dense forests in Siberia. During an interview to BBC, Leonov remembered that he and Belyayev woke up to the sounds of a rescue party approaching:

"They landed 9 km away and came on skis. They made a little hut for us and brought us a big cauldron which we filled with water and put over a fire. Then we washed in it."

For his remarkable success, Leonov was conferred with the prestigious Hero of the Soviet Union Award. He is the lone survivor of the 5 member team associated with the Voskhod space mission. The world is going to always remember him as the Skywalker. Later on he was made the deputy commander of the cosmonaut team to teach them from his experiences on how to perform extravehicular activities.

He received several opportunities to travel back into space including the opportunity to lead the proposed first Russian manned mission to moon and the Salyut 1 mission. While the lunar mission got cancelled after landing of Apollo 11 on the moon, the crew members including Leonov got replaced on the Salyut 1 mission. It was only in 1975 that Leonov got the golden chance to re-enter space and once again on a very special and historic mission Soyuz 19, which was the first joint space project between Soviet Union and United States. In United States the mission was named as the ASTP Project i.e. the Apollo Soyuz Test Project. The mission gave him the opportunity to visit the Johnson Space Center in United States three times and also to learn English. The mission that took off on 15 July 1975, involved docking of the two spacecrafts Soyuz 19 of Russia and Apollo of United States. After docking, the crew on both the spacecrafts changed sides. During this project the two spacecrafts docked twice first for 44 hours and then again for 3 hours. The mission was successfully completed with the safe landing of Soyuz 19 on 21st July 1975. This was Leonov's last and final travel to space. Upon the successful completion of the ASTP, he had said "Dear and intelligent people who decided to show all of humanity that we are different... but can work together."

During his career as a cosmonaut, Leonov stayed for over 7 days in space and received several awards and honours for his remarkable feat in the space. These included both national and international awards like the Hero of the Soviet Union Award (twice), The Order of Lenin, The Order of the Merit to the Fatherland, Medals "For Impeccable Service", Order of Karl Marx, the Order of the Red Star Awards etc. As recognition of his achievements, a crater on the far side of moon has also been named after him as Leonov's Crater. He also received the Honorary Citizenship of countries like Belgorod, Vologda, Kaliningrad, Kaluga, Kemerovo, Nalchik, Perm; Arkalyk (Kazakhstan), Kremenchug (Ukraine), Veliko Tarnovo, Vidin, Svishtov (Bulgaria), Usti and Labem (Czech Republic), San Antonio (Chile) etc. Leonov and his achievements in space have been depicted on several stamps not just in Soviet Union but also in several other countries. IDW publishing has also featured Leonov as a character in the 2013 Doctor Who comic book story "Space Oddity" and a spaceship in the Star Trek Novel Destiny: Gods of Night has been named the U.S.S. Alexey Leonov. Arthur C. Clarke has dedicated his novel 2010: Odyssey Two to Leonov and Andrei Sakharov. Leonov's popularity can also be guessed from the fact that several schools and streets in Soviet Union carry his name. While there is an endless list of the awards and recognitions received by Leonov, it is an interesting fact that his mother was also a recipient of coveted Order of Maternal Glory, which was conferred upon her in recognition of her service as a Soviet woman for the Soviet state by having a nine children.

Before his retirement in 1991 from Soviet Air Force, Leonov was made the Deputy Director of the Yuri Gagarin Cosmonaut Training Center for 6 years (from 1976 to 1982) where he shared his rich experience and knowledge with young cosmonauts preparing them for future missions to explore the universe. Even after his retirement he has served as the vice president of a Moscow bank and was an advisor to the First Deputy of the Board.

Besides being a successful cosmonaut, Leonov was also an accomplished artist, an athlete and a writer. As a first artist in space it is said that he even drew sketches out when in space. Apart from the sketches related to space, Leonov also painted scenes of natural as well as architectural beauty. He has painted over 200 paintings. 'Near the Moon', is one of his famous paintings. "Cyclone over Singapore", is the first cyclone depicted by an artist from space. Some of his other paintings include, "The Ashy Layer", which he painted to depict his theory that there was an "Ashy Layer", in the lower atmosphere that refracts the light of the moon as it passes through, causing a mirage-like reflection on the Earth below. Leonov also painted 'Seagull', in honour of the 25th anniversary of the first woman in space, Valentina Tereshkova, which shows seven seagulls over the earth, representing the seven women who had travelled in space till that time. His works have been widely acknowledged and displayed in the former Soviet Union.

Leonov is a painter of repute too and the sale proceeds of his paintings are often donated for a noble cause. His three paintings, 'Dawn', 'Sunset' and 'Baltic Wave', which he painted with the students of his old were auctioned for 1.5 million rubles. The money was given for the treatment of children with heart disease.

He created an exhibit, called 'Earth and Space Art', a collection of paintings and portraits in which he has painted different parts of the world that he had travelled in recent years. His space works are displayed in Gagarin's Museum and Dresden Gallery, in Tashkent, Almaty, Chile.

Along with astronaut David Scott, Leonov has written a book titled "Our Story of the Cold War Space Race," which depicts the history of the space race between the US and the Soviet Union. He has also written a book with Neil Armstrong. Leonov has also worked for the cosmonaut newsletter APOGEE, as an editor and a cartoonist and another cosmonaut newsletter Neptune as an editor.

In 1984, Leonov founded an association of Space explorers along with his fellow cosmonauts Rusty Schweickart, Vitaly Sevastyanov, and Georgi Grechko.

His life has been full of challenges not just in space but also in his personal life. He dodged death on several occasions including an car accident where his car skidded deep down into a frozen lake along with his wife and an attempt to assassinate him when he was travelling in a car with three other cosmonauts.

Even at the age of 81 years, Leonov leads an active life. He continues to share his experiences and speak at public events. His wife Svetlana is a teacher, and they have two daughters, Viktoria and Oksana.

The following words from Leonov will continue to inspire all those who dream to travel to space and take a look at the earth from the outer space:

"The Earth was small, light blue, and so touchingly alone, our home that must be defended like a holy relic. The Earth was absolutely round. I believe I never knew what the word round meant until I saw Earth from space."

He believes mission to Mars is possible and but would take a definitive decision on the part of world leaders, "One thing is clear: no individual country, like Russia, the US, Ukraine, will do it. This will be an international program. And the crew has to include six people minimum. And there have to be representatives of many countries...."

□

VALENTINA TERESHKOVA
(The First Woman in Space)

Valentina Tereshkova is a former cosmonaut (the Russians called their astronaut as cosmonaut). She is the first woman and also the first civilian to travel to space in 1963, orbiting the earth forty-eight times in Vostok VI. She orbited the earth for almost three days.

Before beginning her career as a cosmonaut, she was a worker in textile-factory assembly and an amateur skydiver as well. Later she became politically active and joined the Communist Party of the Soviet Union and held various political offices, she also travelled the world and promoted Soviet science and feminism. She proved it to everyone that women's qualities and abilities are nowhere less than the men's.

Valentina was born on 06 March 1937 at Volga River village of Maslennikovo in Tutayevsky District, Yaroslavl Oblast, in central Russia. Her full name was Valentina Vladimirovna 'Valya' Tereshkova. Her father Vladimir Tereshkova, was a tractor driver who later served as a soldier in Russian Army during World War II (1939-45) and got killed in the War when Valentina was just two years old. Her mother Elena Fyodorovna Tereshkova was a worker in a textile plant. Valentina is the second amongst her two other siblings. She had a brother named Vladimir and a sister

Ludmilla. She started her schooling when she was eight year old in 1945 but left school in 1953 due to financial reasons. Later in 1945 she moved to her grandmother's home in Yaroslavl, where she started working as a trainee at a tire factory in 1954, she then joined her mother and sister as a loom operator at the cotton mill in 1955.

She was an ambitious and deserving lady. She continued her studies through correspondence and got her graduation degree from "The Light Industry Technical School." As a young girl she was attracted and interested to parachuting. She joined a local DOSAAF Aviation Club and made her first jump at age 22 on 21 May 1959.

She became the secretary of the local Komsomol (A Young Communist League) in 1961, and afterwards joined the Communist Party of the Soviet Union. Valentina Tereshkova had never thought of becoming an Astronaut, her life took a great turn when Sergey Korolyov, the chief Soviet rocket engineer, gave the idea of sending the first woman in space. The main reason behind it was that Soviet Union wanted to set another 'space first', record before the United States. The country also wanted to send a message that the Soviet Union valued all its citizens equally whether men or women. The decision to train a woman as cosmonaut was also intended to motivate little girls throughout the nation, giving them a belief that they, too, could go into space someday.

Valentina was selected to join the female cosmonaut corps on 16 February 1962. Out of almost four hundred applicants, the five shortlisted candidates were: Tatyana Kuznetsova, Irina Solovyova, Zhanna Yorkina, Valentina Ponomaryova, and Tereshkova, among the five she was the least educated woman. The selection procedure involved certain physical and mental standards for selection of the applicant like: they should be parachutists under 30 years of age, under 170 cm (5 feet 7 inches) tall, and under 70 kg in weight. Although Valentina Tereshkova had no prior experience as a pilot, but her experience in skydiving and being a daughter of a war hero made her first and favourite choice of the selection committee as the first women astronaut.

Before her launch in space, she and all other candidates underwent an eighteen months long training. Due to safety and other reasons the whole program was kept a secret. When Tereshkova left for training, she would inform her mother that she was going to a training camp for an elite skydiving team. Her mother came to know about the truth only when she heard the announcement of flight on the radio. The training was very hard and rigorous. All aspiring cosmonauts were trained to live alone for a long period of time, also they were prepared to live in the conditions just like space, with zero gravity pull and weightless body. They did hard practice of parachute jumping, pilot training in MiG-15UTI jet fighters and also learned to use different types of machines required in space. Valentina did excellently in physical but rocket theory and spacecraft engineering were a bit difficult for her to understand.

Once Yuri Gagarin said about her that "It was hard for her to master rocket techniques, study spaceship designs and equipment, but she tackled the job stubbornly and devoted much of her own time to study, pouring over books and notes in the evening."

All the selected candidates were ranked as the second Lieutenants in the Soviet Air Force.

Among the five final candidates Valentina, Irina,

Solovyova and Ponomaryova were the leading once, and it was decided that a joint mission will carry two women in space on solo Vostok flights on consecutive days in March or April 1963. At first it was decided that she would launch first in Vostok 5 and Ponomaryova would follow her into orbit in Vostok 6 but this plan was then re-planned and it was decided that now Vostok 5 would carry a male cosmonaut Valery Bykovsky (who spent five days in orbit aboard Vostok 5 in 1963) followed by joint mission with a woman launch in Vostok 6 in June 1963.

Valentina was nominated as the pilot of Vostok 6 by the State Space Commission at their meeting on 21 May and this news was confirmed by Nikita Khrushchev (a politician who led the Soviet Union during part of the Cold War), her backup was Irina Solovyova. After the successful launch of Vostok 5 on 14 June 1963, she started her final preparation for her flight of Vostok 6. On the morning of 16 June, after her health check-up, her communication and life support checks, Valentina was taken inside the Vostok and was ready for her first flight. After a long countdown of two hour Vostok 6 was launched successfully on 16 June 1963 at 09:29 am from the Baikonur Cosmodrome and Valentina created the history of becoming the first woman in space, during the flight of almost 8 hours, Vostok made 48 orbits of Earth at the altitudes ranging from 180 to 230 kilometers. She was 26 years old at that time.

Chaika was her radio call sign during her Vostok flight. Using her radio call sign she reported, "I see the horizon. A light blue, a beautiful band. This is the Earth. How beautiful it is! All goes well."

Vostok 5 and 6 flied from two different orbits, they both once came briefly together within the range of 5 km of each other, and both the cosmonauts exchanged brief communications through radio. Her journey ended in three days and she landed near Karaganda, Kazakhstan, on June 19, 1963. The photograph taken by her from the space and the flight log maintained by her were later used to identify aerosol layers within the atmosphere. She conducted many biomedical and science experiments during her journey to space to learn about the effects of space on the human body.

After the first flight of a woman in space, second flight of woman took 19 years to go, Svetlana Savitskaya became the second women in space after Tereshkova. After the successful completion of the 'First Woman in Space Mission.' Valentina got married to Andriyan Nikolayev (who was also a cosmonaut) on 3 November 1963 at the Moscow Wedding Palace. The marriage was presided by Nikita Sergeyevich Khrushchev with top government and space program leaders. She gave birth to their daughter Elena Andrianovna Nikolaeva Tereshkova on 8 June 1964, Elena became a doctor and was the first person to have both a mother and father who had travelled into space. Tereshkova and Nikolayev divorced in 1982 after 8 years of marriage. She later married to orthopaedist Yuliy G. Shaposhnikov, who died in 1999.

After her flight Valentina continued her studies and studied at the Zhukovsky Air Force Academy and graduated as a cosmonaut engineer, she also earned a doctorate in engineering in 1977.

She continued to work as an aerospace engineer in the space program. By now she had also become a goodwill ambassador for promoting the equality of the sexes in the Soviet Union and around the world. She became active in Soviet politics and also started working for feminism, and culture. Due to her noticeable personality and her popularity among public, she was chosen to serve for several political positions such as – from 1966 to 1974 she was a member of the Supreme Soviet of the Soviet Union, from 1974 to 1989 a member of the Presidium of the Supreme Soviet and from 1969 to 1991 she was in the Central Committee of the Communist Party. She was retired by the presidential order from the Russian Air Force and the cosmonaut corps in 1997.

Valentina became a face of Soviet Union internationally. In 1966, she was made a member of the World Peace Council, a member of the Yaroslavl Soviet in 1967, a member of the Supreme Soviet of the Soviet Union in 1966-1970 and 1970-974. In UN Conference for the International Women's Year in Mexico City, held in 1975, she was the representative of Soviet Union, where she received a standing ovation. In 1968, Tereshkova headed the Soviet Women's Committee for 19 years.

At the World Conference on women in Copenhagen, she also led the Soviet Union and played a significant role in shaping and promoting the socialist women's global agenda for peace around the world. She holds many important ranks, such as the rank of deputy to the Supreme Soviet, Vice President of the International Woman's Democratic Federation and President of the Soviet-Algerian Friendship Society.

She was awarded with the Hero of the Soviet Union Medal (the medal which is given for heroic feats in service to the Soviet state and society). She was also awarded by many other numerous medals nationally and internationally such as Order of Lenin, Order of the October Revolution, Karl Marx Order etc. A very long list of awards and honours are compiled under her name. She has received United Nations Gold Medal of Peace, Simba International Women's Movement Award, a crater on the moon is also named after her as, 'Tereshkova Crater'. Many schools and streets are also named after her in her honour. She also hold the title of the Hero of Socialist Labour of Czechoslovakia, Hero of Labour of Vietnam, and Hero of Mongoli. In 1997, London-based electronic pop group Komputer released a song entitled Valentina which gives a direct account of her career as a cosmonaut. There are also many short films and other songs too that put light on Valentina's life.

On 5 April 2008, she became a torch-bearer of the 2008 Summer Olympics torch relay in Saint Petersburg, Russia and at the opening ceremony of the 2014 Winter Olympics, she was a carrier of the Olympic Flag.

Valentina is the perfect combination of beauty with brain, she is a courageous, brave, focused, and hardworking woman. Whatever job she was involved in, as a cosmonaut or factory worker or later as a representative at various forums, she did all her jobs efficiently with great responsibility and understanding. She has also worked in raising the standard of women in the society. She wants the women to come forward and enjoy the same rights and equality as the men do. She has always been an advocate of women's rights and space exploration. She is the source of inspiration to everyone especially women, she has seen every face of life from financial problems to relationship problems, but she never gave up, she remained focused on her aim and succeed, she overcame her limits to achieve greatness. There are many books that cover the entire life and journey of Valentina, some of them are "Valentina Tereshkova, The First Lady of Space: In Her Own Words", "It Is I, Sea Gull;

Valentina Tereshkova," by Mitchell R Sharpe, "Into That Silent Sea" by Colin Burgess and Francis French and many other books. Her life journey and struggles are teachings for everyone to keep working for success. She is now 79 years old and still working, she is a member of the State Duma and a loving grandmother. Her youngest grandson Andrey wants to be a pilot, inspired by her.

She still wants to go into the space again. On the occasion of her 70th birthday in 2007, she said, "If I had money, I would enjoy flying to Mars, This was the dream of the first cosmonauts. I wish I could realize it! I am ready to fly without coming back."

□

DENNIS ANTHONY TITO
(The Astronaut Who Funded His Own Trip to Space)

Have you ever dreamt of going to space? About 15 years ago, a US multimillionaire, Dennis Tito turned his dream of going to space into reality at the age of 61 years when he travelled to space and became the world's first space tourist. This dream was the one, he saw as a teenager, after witnessing the launch of Sputnik in 1957. "My dream was to fly in space before I die," Tito said. "And I basically came up with that lifelong goal around the time of Yuri Gagarin's flight."

Dennis Tito was born on 8th August 1940 in Queens, New York. He did his schooling from Forest Hills High School in New York City. Thereafter he went to complete his bachelor's degree in astronautics and aeronautics from New York State University in the year 1962 followed by the master's degree in Engineering from Rensselaer Polytechnic Institute in 1964. After the completion of his studies Tito worked for some time in the Jet Propulsion Laboratory at NASA as an Aerospace Engineer. There he got the opportunity to work for the Mariner 4 and Mariner 9 mission to Mars. He left this job in 1972 to found an investment management and a consulting firm–Wilshire Associates. Making use of the quantitative techniques applied by him at NASA to determine the path of the spacecraft, Tito developed a mathematical model to analyze and determine the risks in the financial markets. He also created the Dow Jones Wilshire 5000 Composite index.

During 1990, Tito was appointed to work on the Board of Commissioners to the Los Angeles Department of Water and Power. There he was instrumental in convincing the board to support the important state ruling pronounced in 1994 for protection of Mono Lake on account of uncontrolled water diversions in the city.

In all these years of working in the financial sector, Tito never lost his connection and love with the mysteries of the outer space. It was this love and passion which led the oldest child of Italian immigrants, Dennis Tito to pursue his dreams and to negotiate the terms and conditions with the Russians Federal Space Agency to take him on a space mission to Space Station Mir in exchange of an exorbitant amount of $ 20 million. Even at the age of 60, Tito was fully determined to travel to space. He had told SPACE.com, "So I was getting over the hill, I thought. So I said, "It's now or never." The decision that received a severe criticism from the Administrator of NASA, Daniel Goldin who was opposed to the idea of allowing a tourist to travel to space. Tito and two of his fellow cosmonauts were even denied training at the Johnson Space Centre because of this very reason.

Unfortunately, after Russia's decision to allow the Mir space station to crash with Earth, the mission got cancelled. However, the his dream remained alive and very soon another company Space adventures Limited entered into an agreement to take him onboard of the Soyuz TM32 mission. The Russians had already agreed to offer Tito a seat on Soyuz but the other station partners including NASA, and the space agencies of Europe, Japan and Canada. Russia was duly informed that they "recommended against", Tito's mission.

According to NASA in principle no objection could be posed for the presence of a paying customer aboard the International Space Station, but Tito's training would be insufficient by April which according to them was the time of complex and crucial station operations.

Tito had to sign a deal stating that in case of a tragedy, none of the national space agencies would be held responsible for that. He also agreed to pay for any breakages caused by him. He was further made to agree that he will not visit the US segments of the ISS without any escort. Tito was trained for a period of one year before the space mission. After completion of training Tito told ABC NEWS that "The training is what counts and I've had a significant amount of training. And that, coupled with my aerospace engineering background, I think puts me in a very strong position, as far as being able to function in space." On 28th April 2001, accompanied by two fellow cosmonauts' commander Talgat Musabayev and flight engineer Yury Baturin, Tito became the first paid tourist to travel to the International Space Station. In an interview transmitted from ISS, the US astronaut James Voss said "We'll welcome anyone who shows up at our hatch, and we'll invite them inside and treat them like we would any spacefarer," Tito spent 7 days, 22 hours, 4 minutes in space and orbited Earth 128 times before parachuting back to Earth with his colleagues on 6th May 2001. After successful completion of the mission, Tito said "For me, it was like being in heaven—it was like being in a second life."

Though he was the first person to fund his own trip to space, he became the 415th person to travel to space at that time. Although he is still remembered as the first space tourist, but Tito himself never liked the term space tourist and had even raised a formal objection to it. Nowadays such travellers, are called the Space flight participants.

After over two years of returning from space Tito submitted his testimony to the Senate Committee on Commerce, Science and Transportation; Subcommittee on Science, Technology, and Space; and the House Committee on Science, Subcommittee on Space and Aeronautics at their joint hearing on "Commercial Human Spaceflight", on July 24, 2003. Dennis Tito promptly returned to his daily routine after his historic flight. He went to work at the position at Wilshire Associates and resumed his daily routine.

"For me, there was very little transition from the space flight to Earth.

Within 48 hours I was actually back to my running – I ran a couple of miles."

He told BBC in an interview on the tenth anniversary of his pioneering space travel.

A few years later in 2013 Tito also revealed his plans of sending a couple, a man and a woman-probably married on a round trip to the planet Mars in 2018.

According to a press release from the Inspiration Mars Foundation, a new organization launched by Tito, "Plans to take advantage of a unique window of opportunity to launch an historic journey to Mars and back in 501 days, starting in January 2018."

It was also revealed that the relatively shorter time period of 501 days for the journey to Mars and back was possible due to a specific alignment of heavenly bodies in January 2018. The same positioning of celestial bodies is not expected to arise again until the year 203.

Tito declared that the mission would be funded by him until 2014 while the remaining financial requirement shall be met through donations from several foundations, private investors and media rights sales. Tito also expects to raise money through selling scientific data to NASA. Stating the urgency of the mission he said:

"There is no time to lose… now is the time."

Inspiration Mars Foundation and its founder Tito hope to inspire a new American Renaissance in science with this vision of theirs which reads: This, "Mission for America", will generate new knowledge, experience and momentum for the next great era of space exploration. It is intended to encourage all Americans to believe again, in doing the hard things that make our nation great, while inspiring youth through Science, Technology, Engineering and Mathematics (STEM) education and motivation.

However, in November 2013, Tito and team members of Inspiration Mars Foundation admitted that their plan was impossible without significant levels of assistance and funding from NASA.

The optimistic mission to Mars is not without considerable hurdles. The size of the capsule in which these amateur astronauts would travel is said to be the size of a toilet. While in transit they can access an "Inflatable habitat module," the module however will detach before entering into the earth's atmosphere. Apart from the uncomfortable space crunch, there is a danger of exposure to harmful radiations that can damage organs, elevate the risk of Alzheimer's disease and cancer.

Another major threat to this mission is from the radiation bursts that the sun sprays out during solar flares or "coronal mass ejections." The mission, although have been planned around the time of low cycling activity of the sun but this only reduces and does not completely eliminate the risk of the spacecraft being flare blasted. NASA had also faced similar risks but these missions were days long not years long.

Tito does not expect to reap riches from this mission, he says:

"Let me be clear, I will come out a lot poorer as a result of this mission but my grandchildren will come out a lot richer for the inspiration it will give them."

After Dennis Tito six other people have had already travelled to space. The journeys however have been very expensive. Tito in an interview to BBC expressed the wish that he hoped that tens of thousands of people can experience what he experienced, for 5% of the cost.

□

GUY LALIBERTE
(Canada's First 'Space Tourist')

Guy Laliberté is an entrepreneur, poker player, philanthropist, stiltwalker, accordion player, fire eater and more but he made a mark in history as Canada's first space tourist. Guy Laliberté was born on September 2, 1959 in Quebec City, Quebec, Canada. When Laliberté was young, his parents took him to the Ringling Brothers and Barnum and Bailey Circus. He was so fascinated by the show that he read up the biography of P.T. Barnum. Guy was always a creative person and e produced several performing arts events while still in school.

After completing his schooling Laliberté decided to pursue his love of show business but started small through giving street performance and playing accordion and harmonica on Quebec streets. He continued performing and learning on the streets and joined Les Échassiers, a performing troupe that hitchhiked across Europe; was only 18 then. The troupe also included fire-breathers, jugglers, and acrobats that from show to show performing all over the country. Laliberté learned the skills of fire eating and stilt walking from his new companions. Guy also attempted to hold on to a full time job later when he returned to Quebec. His new steady full time job was at a hydroelectric dam. But as luck would have it the employees of his new company went on a strike soon after joined work. Laliberté quickly made an exit and returned to the world which he knew better as a successful street performer.

In 1980, he ran as a Rhinoceros Party candidate for the electoral district of Charlevoix in the 1980 federal election. Laliberté quitted his job at the hydroelectric dam and joined a performing group in Baie-Saint-Paul. In the year 1982 helped to organize an international festival for street performers in the town. In 1984 he founded Cirque du Soleil ("Circus of the Sun") that showcased the talent of the buskers from the Fête foraine de Baie-Saint-Paul. Cirque du Soleil was originally created on the behest of the Canadian government as the part of the celebrations to mark the 450th anniversary of Jacques Cartier's discovery of Canada. The government awarded Laliberté a one million dollar contract to develop this celebration. Although, Cirque du Soleil started off as a one-year project but the government of Quebec wanted the show to travel and perform in other provinces too; this kept the show rolling. Now an internationally renowned circus, Cirque du Soleil was cofounded by Laliberté and Gilles Ste-Croix. It is said that Laliberté is the 'Father' of Cirque du Soleil, while Ste-Croix is the 'grandfather'.

Laliberté had thought of the name 'Cirque du Soleil' or 'Circus of the Sun', while holidaying in Hawaii and says that the tag reflects his vision of the show that, "the sun stands for energy and youth" and that the circus is about those two words. The circus had a tough start when the tent collapsed on the very first day of the show but finally became a huge hit with people and was a financial success too. Since 1984, Laliberté has been an integral part of every production of the Cirque du Soleil shows. The company is one the biggest and most watched live entertainment groups across the world. The Cirque has served its visual treats for more than 160 million people across more than 400 cities on six continents. Shows from Cirque du Soleil are known for its state of the art make up, activities, stage performances and the influence it leaves on the audience. The shows make use of approximately 4,000 people coming from nearly 40 countries of the world and generate estimated annual revenue of more than exceeding 810 million US dollars.

As per the March 2012 figures net worth of Laliberté wealth was estimated at 2.6 billion US dollars following which *Forbes* magazine ranked him as the wealthiest man at 11th rung in Canada and at 459th position in the world. An official statement released on April 20, 2015 from the Cirque du Soleil stated that Laliberté sold major part of his business (90% stake) to a U.S. private equity firm TPG Capital and Chinese investment group Fosun while keeping a 10% stake in the company with himself. Even after the change in ownership Laliberté continues to provide strategic and creative input to the company.

In 2009 Laliberté became Canada's first space tourist. During the news conference conducted in Moscow, where he confirmed his plans of space travel, he shared that he has long been fascinated by space travel and that "the trip is an expression of a childhood dream." The news conference was telecast live to the Canadian Space Agency in Longueuil, near Montreal.

"When man walked on the moon for the first time, I realized the book, The Little Prince could become a reality," he said, alluding to the famous work of the French aristocrat, writer, poet, and pioneering aviator Antoine de Saint-Exupéry, in the same conference.

The year 2009, in which Laliberté made his self-funded

space travel, marks the 25th anniversary of Cirque du Soleil, the 20th anniversary of the Canadian Space Agency and, in September of the same year, Laliberté's 50th birthday. He called his September 2009 space trip a case of "planetary alignment."

Laliberté also tried bringing fun into this journey and officially became the first clown in the space with his paid space trip. The entertainer donned a bulbous red clown nose before boarding the spaceship and packed several of these novelty noses for other astronauts and cosmonauts waiting for him at the International Space Station.

Laliberté took off from the Baikonur Launch Facility located in the Kazakhstan steppe aboard the Soyuz Spacecraft along with a Russian cosmonaut Maxim Survayev and an American astronaut Jeffrey Williams. The earlier six space tourists had booked their space trips through Space Adventures of Virginia and paid $35-million for a round-trip ticket to the International Space Station. Laliberté did not reveal the exact figure of the amount involved because of a confidentiality agreement but he said it was "pretty similar", to those that were earlier negotiated. The money went to the Russian space agency.

The Quebec billionaire made his new home 220 miles (354 km) above Earth for 12 days and made friends with the other eight space flyers aboard the station. He also got to meet Canadian astronaut Bob Thirsk, who was on a six-month visit to the International space station.

During his 12 days stay aboard the International Space Station Laliberté took several pictures of the Earth's endangered water systems; the limited numbers large-scale prints of which were auctioned by him to raise money for the initiative One Drop. Laliberté had officially launched

One Drop Foundation on October 29, 2007 to help make safe water accessible around the world. Some of the water bodies photographed by Laliberté were Lake Duli Shihu in Tibet; Kazakhstan's Lake Balkhash, the second-largest lake in Central Asia, with a surface area of more than 16,000 square kilometers; the Qaidam Basin located in Haixi Mongol and Tibetan Autonomous Prefecture in China; the Gulf of Mexico, the ninth-largest body of water in the world and also known as the "Mediterranean of the Americas. Laliberté said in an interview that he hoped to use the self-funded trip to space to raise awareness about the need for clean water around the world. Laliberté's trip culminated October 9, 2009, when also held an artistic mission to support his cause of spreading awareness about shortage of clean water". The mission was called "Moving stars and earth for water," and held at the same time across 14 cities around the world and out of the world on the International Space Station. Participants in the mission looked up at the sky for 125 minutes; looking up at the stars to pay tribute to the amazing natural resource called water.

In a video-link shared by Laliberté while aboard the International Space Station, he said about his space trip, "Yes, it's worth every penny...," A smiling Laliberte in the video also said that he is getting used to floating in weightlessness and the trip is more than worth the multimillion-dollar price tag. He further added that weightlessness has been a joy even though he did smack his head on the ceiling three times in one day.

After his return from the space in June 2011, Laliberté published a book, titled Gaia, which contains photos of Earth from his 2009 trip to the International Space Station. The sales proceeds from this book were donated to the One

Drop Foundation.

Apart from being a busker and founder of a successful international circus Laliberté is also an avid poker player. He finished fourth in the World Poker Tour Season Five event at Bellagio in Las Vegas and won $696,220 in April 2007. Other poker shows in which Laliberté played include GSN's High Stakes Poker Season 4 show and Poker After Dark Season 4. He was also known for frequenting the highest stakes games on Full Tilt Poker. Laliberté announced the Big One in 2011. The Big One was a 1 million dollar buy-in tournament and was also featured at the 2012 World Series of Poker. Part of the prize pool created out of the buy-in was given away as charity to Laliberté's philanthropic organization One Drop Foundation. Antonio Esfandiari won the tournament over 48 other players with the prize money $18,346,673. A whooping sum of $5,333,328 dollars was raised for the One Drop Foundation through the tournament.

Laliberté took inspiration from Cirque du Soleil and Cirque du Monde (the international program for street children), and used the creative experience to design the functioning of the One Drop Foundation. The project was launched with a view of fighting world poverty through making clean water available and accessible to everyone. The foundation works through performing circus arts, popular theatre, folklore, dance, music, and the visual arts to promote education, community involvement and public awareness of water issues. The technical projects of the foundation are aimed at improving access to water, ensure food security and promote gender equality in communities especially in developing countries.

Laliberté has arranged for covering the operating costs One Drop through the 100 million contributions to be paid over a period of 25 years. Field activities organized the foundation are financed through the public donations, donations by the employees of Cirque du Soleil and through the funds allocated by Canadian and international partners.

In an interview with Sapce.com, an online resource for space lovers and professionals alike.

On being asked why according to him it was important for artists and non-astronauts to go to space along with the professionals he said, "...Well, I think this is an amazing experience that I wish more people could experience, in terms of being able to look at this planet from a point of view that shows its beauty in such a large scale. It's very emotional, very touching. You don't have to be an artist to feel like that. Every cosmonaut or astronaut I met there always has said, "We all have that thing in common, there's the very deeply emotional feeling that you live when you're in space." In the end, it does have an impact on the way you see Earth, the way you see life. If there was more people able to live that, maybe we'd have a better planet."

In the same interview when asked whether he would consider going back to space, he answered that he was busy in several projects at present and believed that no seat would be available so soon again for the space travel. In a lighter vein he added, "...Still, I keep my door open for an alien ship to come and pick me up and bring me farther, but that we'll see."

□

GUION BLUFORD
(The First African American in Space)

Guion Bluford is a former NASA Astronaut. He became the first African-American to be a member of a space flight when he flew on the spaceshuttle Challenger in 1983 and second person of African origin in space, after Cuban cosmonaut Arnaldo Tamayo Méndez. Incidentally, Bluford's first space mission was also responsible for deploying India's artificial satellite in space. Before serving as an astronaut, he has also served as an officer to Colonel in the U.S. Air Force. Guion Bluford had been to space four times, spending near about 28 days 16 hours and 33 minutes there. After his retirement he became a businessman.

His complete name was Guion Stewart Bluford, Jr. Now 74, was born on 22 November 1942 in West Philadelphia, Pennsylvania. Everyone in his family was well-educated. His father Guion, Sr. was a mechanical engineer and mother Lolita, was a special education teacher. Blurod parents motivated him to work hard and also to set high goals. Bluford had two brothers Eugene and Kenneth. Although not very bright academically, Guion was a hardworking and sincere boy from his childhood. He was also very quiet and liked to remain private.

Bluford's brother, Kenneth said in an interview that "He just had to work harder than the rest of us. He put in very long hours. He was always a little behind and trying to catch up. He was not like a kid who was unusually bright, with his mind darting all over the place, making discoveries here and there. In school, he was always slugging it out. He liked to build model airplanes and put things together in his spare time."

He was popular among his family and friends with the nicknames 'Bunny' and 'Guy'. In his early years, he participated in Boy Scouts and later became an Eagle Scout. He did his schooling from Overbrook Senior High School in 1960, where he was told by his teacher that he must join some technical college. In 1964, he received his Bachelor of Science degree in Aerospace Engineering from the Pennsylvania State University as a distinguished Air Force ROTC graduate. He completed his Master of Science degree in Aerospace Engineering from U.S. Air Force Institute of Technology in 1974. Later he also got a Doctor of Philosophy degree in Aerospace Engineering from the same college in 1978. Guion earned a Master of Business Administration degree from the University of Houston–Clear Lake in 1987. He also attended the Wharton School of Business of the University of Pennsylvania.

One of his professor of aerospace engineering at Penn State, told a newspaper that Bluford was, "a quiet fellow and an average student, not the sort you would expect to be interviewed about some years later."

His interests other than flying includes reading, swimming, jogging, racquetball, handball, scuba diving and golf. Married Linda Tull of Philadelphia, Pennsylvania in 1964, while he was in his senior years. They have two sons Guion Stewart III, born on June 12, 1964 and James Trevor, born on October 25, 1965. Bluford joined pilot training at Williams Air Force Base formerly known as the United States Air Force (USAF) base in Mesa, Arizona, on January 1966 he received his pilot wings because of his hard work and dedication towards flying. He then did his F-4C combat crew training in Arizona and Florida and was assigned to the 557th Tactical Fighter Squadron, Cam Ranh Bay, Vietnam. Bluford flew near about 144 combat missions as a fighter pilot in the Air Force. Before being selected by NASA Bluford had already accomplished many milestones in his career. In 1967, Bluford started working as a flight instructor at the 3630th Flying Training Wing, Sheppard Air Force Base, in Texas; by 1971, he became the executive support officer to the deputy commander of operations and also the school secretary for the Flying Training Wing. During his association with the Air Force Flight Dynamics Laboratory at Wright-Patterson Air Force Base in Ohio, Guion learned, researched, wrote, and presented many scientific papers on computational fluid dynamics. He logged about 5,200 hours in jet flight and also acquired an FAA (Federal Aviation Administration), commercial pilot license. He is also a certified scuba diver. Guion Bluford also fought in the Vietnam War.

Bluford was among 35 candidates chosen by National Aeronautics and Space Administration's (NASA) to join their new spaceshuttle team in January 1978 out of some 10,000 applicants to the space program, He became a NASA astronaut officially in August 1979. Here he was given the technical assignments that included working with the Remote Manipulator System (RMS), Space Station operations, Spacelab systems and Space Shuttle systems. His duties with NASA also included looking after payload safety issues and verifying flight software in the Shuttle Avionics

Integration Laboratory (SAIL) and in the Flight Systems Laboratory (FSL). He has also been a mission specialist on many flights like STS-8, STS-61-A, STS-39, and STS-53.

Blurford's first mission was on STS-8 shuttle Challenger that was launched from Kennedy Space Center, Florida, on August 30, 1983. The mission was the first night launch and first night landing mission. Bluford was the mission specialist in his first flight, it was the third flight of the Challenger. STS-8 did 98 orbits of the Earth in 145 hours and Bluford with his crew members conducted many experiments, they also operated a Canadian-built robot arm during the mission. The STS-8 crew deployed the Indian National Satellite (INSAT-1B) in space. During the mission the crew also operated the Continuous Flow Electrophoresis System with live cell samples; and conducted medical measurements to understand biophysiological effects of space flight. The mission ended successfully on September 5, 1983, when the spacecraft landed down safely at Edwards Air Force Base in California in a night landing.

After his first flight Bluford said, "It might be a bad thing [to be first], if you stop and think about it," Bluford told the *Washington Post* (a newspaper), "It might be better to be second or third because then you can enjoy it and disappear—return to the society you came out of without someone always poking you in the side and saying you were first."

After his first and remarkable flight to space Bluford became an international hero and a role model for many people of African origin, who were marginalized even in the modern and liberal American society. His success proved that colour or creed can't define the success nor they be an obstacle in the way to the success.

Two years after making the history, Bluford made his second trip to space as a specialist for mission STS 61-A aboard Challenger on October 30, 1985, that was the last successful flight of the spaceshuttle, after this mission the shuttle exploded on its next flight in 1986, in which seven crew members including the first teacher in space, Christa, were killed. Mission STS 61-A hold the record of the largest crew aboard any single spacecraft for the entire period from launch to landing (eight people); The crew members included Henry W. Hartsfield, Jr., commander; Steven R. Nagel, pilot; Bonnie J. Dunbar, James F. Buchli and Guion S. Bluford, mission specialists; and Ernst Messerschmid and Reinhard Furrer of West Germany, along with first Dutch astronaut Wubbo Ockels of the European Space Agency (ESA), all payload specialists. It was a scientific Spacelab mission, funded and directed by West Germany, the Payload operations of this were controlled from the German Space Operations Center in Oberpfaffenhofen, West Germany, instead of from the regular NASA control centers. The flight took 11 orbits year in 169 hours, and landed (Challenger) safely at Edwards Air Force Base on November 6, 1985.

After his second flight, Bluford was grounded for some time due to a herniated disc (often referred to as a slipped disc), just 4 months before his third flight. Remembering the hard time he said, "The NASA flight surgeons grounded me and indicated that I would need an operation to correct the problem," He recalled, "There was some concern from the training folks that I might not be able to complete the training syllabus in time for the flight. Commander Mike Coats altered some of the responsibilities on the crew, so that I could be operated on and still make the flight." Bluford with his efforts and determination overcame the difficulty and finally made his third flight mission STS-39 the orbiter Discovery, which was launched from Kennedy Space Center on April 28, 1991, he performed various experiments with the crew for the U.S. Department of Defense, and the space-shuttle completed 134 orbits in 199 hours and landed on May 6, 1991 at the Kennedy Space Center.

Bluford's final trip to space was on December 2, 1992, as one of five crew members of mission STS-53 aboard Discovery. The crew included Commander David M. Walker, Pilot Robert D. Cabana, Mission Specialist Michael R. Clifford, Mission Specialist James S. Voss and Bluford himself as mission specialist. Discovery carried a classified payload for the Department of Defence, two unclassified secondary payloads and nine unclassified middeck experiments. The crew performed several Military-Man-in-Space and NASA experiments. The spaceshuttle completed 115 orbits of the Earth in 175 hours and landed successfully at Edwards Air Force Base on December 9, 1992. With four successful accomplishments, Bluford has logged about 688 hours in space in total. After landing from its fourth flight, Bluford said he "had to seriously decide", what to do next, and he realized that it was time for him to leave NASA and take retirement. Bluford left NASA and also took retirement from the Air Force in July 1993.

After his retirement he took the post of Vice President/ General Manager, Engineering Services Division of NYMA, and Greenbelt, Maryland. He got engaged with many private sectors and worked with them in various positions. He became the Vice President of the Aerospace Sector of Federal Data Corporation in May 1997. In 2000 October Guion Bluford took up the position of the Vice President of Microgravity R&D and Operations for the Northrop

Grumman Corporation. In 2002 he became the President of Aerospace Technology, Cleveland, Ohio an engineering consulting organization. Despite his retirement from NASA, he maintained ties with the organization for some of the space program. One such notable program was when he worked with the Columbia Accident Investigation Board that investigated the unfortunate and sudden break-up of spaceshuttle Columbia in February 01, 2003 killing all of its seven crew members.

Bluford is also associated with many organizations as an active member including American Institute of Aeronautics and Astronautics; Board of Governors, National Space Club (1997 to 2001); he has also been in the board of directors of many organization such as the Western Reserve Historical Society, the Great Lakes Science Centre, National Research Council (NRC) Aeronautics and Space Engineering Board (1993 to 1998). He is also very active in social activities and he was in the board of directors of society of Distinguished Alumni, Pennsylvania State University, Committee on Minority Activities and many more.

For his remarkable career and his unforgettable achievements, he has been awarded with numerous awards and honours such as Air Force Institute of Technology Distinguished Alumni Award, The Pennsylvania Society Gold Medal, 1991 Black Engineer of the Year Award, National Society of Black Engineers Distinguished National Scientist Award (1979), U.S. Astronaut Hall of Fame Inductee and many more. He was also listed among 100 Greatest African Americans by Molefi Kete Asante in 2002.

On being asked about his space career he shared that "I wanted to set the standard, do the best job possible so that other people would be comfortable with African-Americans flying in space and African-Americans would be proud of being participants in the space program and encourage others to do the same."

Bluford is quiet and simple nature and he is true combination of courage, dedication, hard work and enthusiasm to learn and to do new things. An American Journalist said that "Bluford's career" proves "That the right stuff' comes in hues other than white."

□

EILEEN COLLINS
(The First Female Pilot and Commander of a Space Shuttle)

Eileen Collins is a retired American astronaut and also a retired Colonel of United State Air Force. She has registered herself in history by achieving many "Firsts", under her name in space travel. In 1995 she became the first woman to pilot a spaceshuttle; in 1999, the first woman to command a spaceshuttle, (only two women in the history had commanded spaceshuttle–the other is Pamela Melroy). In July 2005 she commanded the much-awaited launch of the first spaceshuttle after the disastrous Columbia voyage in 2003 (in which all the seven crew members died). She has achieved numerous awards and honours, and is remembered as a courageous, fearless, brave and strong astronaut. She has experienced four space flights and has logged 38 days 8 hours and 10 minutes in outer space. Since her retirement in May 2006, she is pursuing her personal interests and likes to spend more time with her family. Collins is also serving as a board member of USAA (United Services Automobile Association).

One of the NASA Chief Michael Griffin, once said in a statement that "Eileen Collins is a living, breathing example of the best that our nation has to offer, she is of course, a brave, superb pilot and a magnificent crew commander."

She is an energetic and thin woman, with short, curly hair and friendly, lovely brown eyes.

Eileen was born to be a star on November 19, 1956 at Elmira, New York, the most suitable birthplace for a future star pilot to born, because the city is known as the, "soaring capital", of the United States. Harris Hill Soaring Center in Elmira is a renowned place where pilots come together to fly gliders. Her father James E. and mother Rose Marie got separated when she was just nine years old. She has a brother named Clark, and two sisters named Amanda Collins, and Emily Collins. From her childhood, Eileen had a special love and interest for flights and airplanes. She always wanted to be a pilot. When she was a child her favourite hobby was to go to the airport with her parents where she liked to watch the takeoff and landing of the planes while she sat in the car. After the separation of her parents, she faced an emotional as well as an economical downfall in her life and due to this her need and wish to take flying lessons suffered but Eileen never compromised with her dream and she never gave up. She was very determined and focused to be a pilot. When she was in high school she worked at a pizza parlor at night to save the 1,000 dollars that were needed for taking private flying lessons.

She once said in an interview, "When I was a child, I dreamed about space -- I admired pilots, astronauts, and I've admired explorers of all kinds. It was only a dream that I would someday be one of them. It is my hope that all children, boys and girls, will see this mission (her second mission as a commander of the spaceshuttle) and be inspired to reach for their dreams, because dreams do come true!"

At the age of 19 she stepped on her first plane and

that was the time when she realized that she would be a professional pilot one day. She also continued to read and learn about military flying. She graduated from Elmira Free Academy, Elmira, New York, in 1974. Her first step towards the military was when she joined the Reserve Officer Training Corps (ROTC). The college-based program under ROTC was focused at preparing individuals for advanced military careers. The scholarship grant under ROTC helped her to attend Syracuse University in New York from where she earned her Bachelor of Arts degree in mathematics and economics in 1978. She completed her Master of Science degree in Operations Research from Stanford University in 1986; and a Master of Arts degree in Space Systems Management from Webster University in 1989. She was one of the four women chosen for undergraduate pilot training at Vance Air Force Base, Oklahoma out of the 120 women applicants. From the total strength of 320 students only four (including her) were women. After her one year training and after getting her flying wings she stayed at Vance for three years as a T-38 Talon instructor pilot. From 1979 until 1990 Collins taught flying at bases in Oklahoma, California, and Colorado. She also worked as an assistant professor of mathematics at the U.S. Air Force Academy in Colorado Springs. Eileen was the second female pilot to attend the U.S. Air Force Test Pilot School and graduated in 1990 following which National Aeronautics and Space Administration (NASA) selected her for the astronaut program. In 1987 she married her fellow pilot Pat Youngs they had met in the 1980s as flight instructors in California. They have two children, Bridget and Luke. Other than flying she also enjoys running, golf, hiking, camping, reading, photography and astronomy in her spare time. After her selection as the pilot of spaceshuttle, Eileen underwent basic astronaut training. Which included many courses in land and water survival tactics, parachute training, and field trips to various NASA centers and geological sites. Eileen and rest of the astronauts were trained to live in the similar environment as are expected in space with zero gravity and other factors. They also attended classes for history of the space programs, weather, medicine, and mechanics. They were taught about the formation of spaceshuttle and how to operate it, but the most difficult part of training was the simulator. In the simulator pilots are put through the practice launches of the shuttle. The session is eight-minute long during which instructors bombard trainees with a series of mechanical malfunctions that may occur during the actual launch or afterwards in the mission (as happened with Columbia in 2003). The pilots were given a few seconds to make a life-or-death decision and to act and to save their as well as the life of the crew. Eileen's initial assignments and work were to provide engineering support for unmanned orbiting systems. Over the years she also served as a spacecraft communicator, and then as the Astronaut Office Spacecraft Systems Branch Chief, Chief Information Officer, Shuttle Branch Chief, and Astronaut Safety Branch Chief. Her long training and various profiles as well as the astronaut practice prepared her for her first mission as a spaceshuttle pilot.

Eileen first flew the Space Shuttle as a pilot in 1995 aboard STS-63. The mission was a deal between Discovery and MIR (the Russian space station). On remembering her first flight she says':

"The launch sounds like you're standing in a room that's

on fire. The engines turn off at eight and a half minutes, and you're immediately in zero gravity. I pulled out my pen and it floated. I thought, I'm here—I'm in space."

She received the Harmon Trophy. Which is actually a set of three international trophies, to be awarded annually to the world's outstanding (aviator, aviatrix, and aeronaut). The trophy was given to her for her extra- ordinary work and in recognition of her being the first female Shuttle Pilot. Later she also piloted STS-84 in 1997.

Eileen again made history on July 23, 1999 when she became the first female commander of a U.S. Spacecraft Shuttle Mission STS-93. The mission deployed the Chandra X-Ray Observatory. Through the Chandra X-Ray Observatory astronauts observe very hot regions such as exploded stars, clusters of galaxies, matters around black holes around the Universe for X-ray emissions. She faced several problems through the mission including the backup engine glitch, fuel line leakage. However, throughout the ordeal Commander Eileen remained calm and patiently guided the spacecraft through the launch and its five-day mission. She also made a successful landing despite of all the unforeseen troubles during the mission.

Eileen was now a famous and respected name in her profession but she earned media attention just before her retirement from NASA in July 2005. Her last mission with NASA was to command the launch of the first spaceshuttle after the tragic incidence of spaceshuttle Columbia in 2003. Columbia, upon its re-entry in the earth's atmosphere, went up into flames killing all seven astronauts onboard including the first Indian origin woman astronaut Kalpana Chawla. Her 2005 mission was looked upon with anticipation by NASA as it held the key to the future of manned space flights. The mission focused on testing out the new safety measures and repair techniques for the future missions as well as to ensure the safety of both the astronaut and space shuttles.

Eileen was asked before her flight if she feared for her safety or if she felt pressured that the future of manned spacef light depended on her success.

To this her calm reply was, “We are staying focused on the mission and we know we are in good hands with the people on the ground.”

Eileen and other crew members stayed in space for fourteen days. The crew spent most of the time docked at the International Space Station (ISS). ISS is sponsored by six international agencies from the United States, Japan, Russia, Canada, Brazil, and members of the European Space Agency as an orbiting station. The crew jobs included delivery of supplies to the ISS and making routine inspections and maintenance. In addition, Eileen and other crew members carefully examined the shuttle to ensuring that no damage was borne on their journey towards the space and the shuttle was in good condition to give them a safe landing on the earth. The shuttle returned safely and made a successful landing on the earth on August 10, 2005 after a brief weather delay; all the seven crew members, ground officials at NASA watchers all-around the world were relieved.

Program manager Bill Parson at a CNN news conference, said, “It’s a good day to be us.”

At the same conference, another senior official praised Eileen for the success: “There isn’t any of this that is easy... but Eileen made it look like a cakewalk.”

On May 1 in 2006, Eileen announced her desire to

leave NASA to spend more time with her family and to purse her personal interests. In her announcement she said "It has been wonderful," but the number one thing for me now is to spend time with my family.

Her hometown Elmira honoured and awarded her on several occasions celebrating every milestone and feat of her career. She has achieved almost an idealized status in Elmira inspiring every woman and young girl there. An Eileen Collins Observatory is established at Elmira under the wings of Corning Community College.

Her honours and awards include; Distinguished Flying Cross, the NASA Outstanding Leadership Medal and NASA Space Flight Medals to name a few. Eileen was included in the National Women's Hall of Fame in Seneca Falls, New York and thus joined the ranks of America's top female achievers after her first space flight in 1995. Eileen has also received the Free Spirit Award, and the 2006 National Space Trophy. She is an international celebrity but she never seeks limelight. Her motto is to lead an ordinary life with high values. She has always remained very private about personal life and keeping her family out of the limelight. Encyclopedia Britannica recognizes her amongst the top 300 women in the history who have changed the world.

Eileen was awarded Douglas S. Morrow Public Outreach Award in 2007. The award is given annually to an individual or the organizations, which had significantly contributed towards creating public awareness about space programs.

Syracuse Hancock International Airport has named its main entrance boulevard in her honour. She is associated with many organizations in various capacities. She is the member of the Air Force Association, Order of Daedalians; Women

Military Aviators, U.S. Space Foundation; the American Institute of Aeronautics and Astronautics, and the Ninety-Nines–An International Organization of Women Pilots.

Eileen Collins's life has a lot to teach and inspire everyone. A book titled "Eileen Collins", by Elizabeth Raum was published in 2005. It describes the life history of Eileen, and traces her inspiring journey of life from an ordinary girl to a star astronaut.

□

4
Mishaps in Conquest of Outer Space

It's not just the desire of human beings to explore the outer space but also the race among several nations to exhibit their technological supremacy in the outer space that has been behind the numerous space missions undertaken till date. While the success of these missions has helped in unveiling of many secrets of the Universe, the failures or mishaps on some of these missions have just inspired the teams to come back stronger the more prepared to face the unknown challenges. Many of the so- called successful missions also were not free from the nail biting moments where not just the success of the mission but also the life of the crew members was at stake. Given below is the narration of the unforgettable mishaps that occurred on a few of such missions.

Voskhod 2

"My suit was becoming deformed. My hands had slipped out of the gloves [and] my feet came out of the boots. The suit felt loose around my body. I had to do something. I couldn't pull myself back using the cord. And what's more, with this misshapen suit, it would be impossible to fit through the airlock."

These are some of the reflections of Alexey Leonov, the first astronaut to successfully carryout Spacewalk. It was on 18th March, 1965 that Alexey Leonov begun his journey along with his fellow astronaut Pavel Belyayev aboard the Voskhod 2 spacecraft. Voskhod 2 was launched from the Baikonur Cosmodrome in Kazakhstan. Leonov spent an exhilarating 12 minutes outside the spacecraft as he accomplished the mission of first ever spacewalk. After the initial hiccups with his suit during the spacewalk everything appeared to be in control until the spacecraft began its re-entry to land on earth when its automatic orientation device malfunctioned, forcing Voskhod 2 to land manually in a dense forest far away from the expected location. Both the astronauts had to spend hours in that dense forest in freezing temperatures and fear of being attacked by wild animals, before the rescue team spotted them and brought them back in a helicopter.

Gemini 8

Gemini 8 was the sixth manned mission under the Gemini program launched by NASA in its preparation to touch the surface of moon. It was also the 22nd manned space flights of all times. The primary objective of the Gemini 8 mission was to perform the first ever docking with an unmanned spacecraft in the orbit. Gemini 8 was successfully launched on 16 March, 1966 from Cape Canaveral, Florida just a little after the launch of unmanned spacecraft Agena. This unmanned target vehicle is also popularly known as the 'Gemini Agena Target Vehicle', or 'GATV'. Within six hours after the launch, astronauts Neil Armstrong and Dave Scott successfully accomplished the mission's objective of docking with Agena in the earth's orbit. However, soon after the successful docking, the two spacecrafts started to roll and spin. Without losing his calm in this life threatening situation, Neil Armstrong tried to control the situation by first balancing the Gemini 8 and then later on by swiftly undocking with Agena. However, he soon realized that the problem was not because of malfunctioning of systems on Agena but because of malfunctioning of thrusters on Gemini 8 only. In spite of being in a situation which could have killed him much before he landed on moon, Armstrong managed to switch off the trouble creating thrusters of Gemini 8 and initiate an early re-entry. The spacecraft successfully landed somewhere in China within 11 hours from its launch and both the astronauts returned safely.

Apollo 1

The ill-fated mishap of Apollo 1 occurred when at 1:00 p.m on January 27, 1967, the three astronauts Virgil "Gus" Grissom, Edward H. White II, and Roger B. Chaffee entered the Apollo 1 command module mounted on an unfuelled Saturn Rocket for the purpose of training for a future launch through rehearsal of the countdown sequence. The incident occurred at Cape Canaveral, Florida. As soon as the crew entered the module, they started experiencing the problems. The stimulated countdown was put on hold at 6.30 p.m. at T minus 10 minutes and was to resume within one minute. It is at this moment that the crew inside reported of fire inside the cockpit and within a matter of few seconds, all the three crew members inside the spacecraft were reported dead. Though the reasons of the fire remain debated and officially unexplained but it is known that the ill-fated spacecraft had a history of technical problems in the past. The disaster not only caused loss of three precious lives and a huge financial dent but also brought down the morale of the entire team associated with the program. However, undeterred by the irreparable loss, the team worked together to correct the deficiencies in the spacecraft.

Soyuz 1

As a precursor to the planned lunar landing mission, with Soyuz 1 mission, Soviet Union wanted to show supremacy over US in Space Program after the recent failure of Apollo 1 mission. Soyuz 1 was manned by Colonel Vladimir Komarov. Just after the launch of Soyuz 1 it was planned to launch Soyuz 2 on the very next day with Bykovsky, Khrunov, and Yeliseyev. It was planned that Khrunov and Yeliseyev would spacewalk to Soyuz 1 and bring back Komarov to earth. Unfortunately, things didn't proceed the way they were planned and soon after the tragedy of Apollo 1, the world witnessed yet another tragedy with the death of Colonel Vladimir Komarov. As Komarov began his journey on April 23, 1967 abroad the new Soyuz spacecraft and entered the orbit, nobody could have imagined that this journey was going to end on such a tragic note. As the spacecraft travelled from the 1st to the 19th orbit, Komarov faced a series of problems including the failure of solar panel to open and loss of power to critically low levels forcing him to attempt a manual re-entry on the 19th orbit. In spite of his desperate efforts to deploy the main and the backup parachutes, the spacecraft failed to deaccelerate and the craft came hurtling down on the ground in the southern Ural Mountains. Komarov could not survive and the dream of Soviet Union also suffered a setback.

Soyuz 11

In a mission to send the crew members to board the space station Salyut 1, on 30th June 1971 Soviet Union launched The Soyuz 11 with three crew members Georgi Dobrovolski, Viktor Patsayev, and Vladislav Volkov. The crew was supposed to stay at the space station and carry out various studies and experiments for a period of 30 days before returning to earth. Breaking the previous space endurance record of 18 days, the three crew members successfully stayed onboard Salyut 1 for 22 days and performed several important tasks including observing human response to the prolonged weightlessness. After the landing of Apollo 1 on the surface of the moon on 20th July 1969 by United States, the success of this mission of manning a space station for the first time was very crucial for the Russians. While the Russians were celebrating and waiting for the return of their heroes, on June 29th, 1971 the crew members started their return journey onboard Soyuz 11. As the spacecraft entered the Earth's atmosphere, a failure in the firing of the pyrotechnic devices caused a sudden pressure drop inside the module. The spacecraft landed successfully, however the three heroes of the mission lost their lives due to decompression and lack of oxygen inside the cabin. It is said that their lives could have been saved if they were given the space suits. As a learning from the failure, since then, all the cosmonauts are given specially designed suits which are required to worn in all stages of the mission which may experience be decompression of the spacecraft.

Soyuz 23

What happens when a space mission planned for at least 73 to 85 days has to be aborted within a day after the launch? This is exactly what happened with the Russian spacecraft the Soyuz 23 which was launched on 14th October 1976 and had to return to Earth just after a day on

16th October 1976. As per the plan, the spacecraft carrying the astronauts Vyacheslav Zudov and Valery Rozhdestvensky was to dock with Salyut 5 and stay there for the planned period. However, on 15th October 1976 as the spacecraft tried docking with the space station Salyut 5, the docking system failed and the spacecraft started moving away from Salyut 5. The launch was plagued with problems from the start. First, the bus transporting the cosmonauts to the launch pad brokedown. Then, at liftoff, the craft began to veer off-course due to heavy winds at the launch site. Once in orbit, Soyuz tried docking with Salyut 5, but the docking program malfunctioned and turned them away from the space station. After all efforts to rectify the problem failed, it was decided to bring back the spacecraft to earth. Soyuz 23 landed on earth at 8.45 p.m. on 16th October 1976 but instead of landing in Arkalyk Kazakhstan, the spacecraft landed about 75 miles away on a freezing lake, Tengiz in rough weather conditions of a blizzard, dense fog and a temperature of −22°C. Soon after their landing the parachutes got filled with water which dragged down the capsule below the surface of lake making it further difficult for the rescue team to locate them. It was only the next day that the capsule could be dragged to the shore and both the astronauts were fortunate to come out safe.

Space Shuttle Challenger

Just 73 seconds after the liftoff from Kennedy Space Center, On January 28, 1986, the American Space Shuttle, Challenger broke up marking a disastrous end to the 10th mission of the Spacecraft. NASA suffered its most devastating disaster. Within no time, the seven crew member onboard the Space Shuttle Challenger—Greg Jarvis, Christa McAuliffe,

Ronald McNair, Ellison Onizuka, Judith Resnik, Michael J. Smith, and Dick Scobee—lost their lives as people across the world witnessed the shuttle suddenly bursting into flames live on their television screens. Those who lost their lives included an ordinary school teacher Christa McAuliffe who was slated to be the first ordinary American citizen to go to space. The reason for this devastating failure has been attributed to the failure of an O-ring due to cold weather. The spacecraft was originally scheduled to be launched six days earlier but it got delayed on account of bad weather and some technical issues. The first journey of the spaceshuttle Challenger was made three years earlier to this accident on 4th April 1983. After Columbia, Challenger was NASA's second reusable manned spacecraft. After the catastrophic disaster, NASA had suspended its space programs for almost two years. The program was resumed in 1986 with the launch of the spaceshuttle Discovery.

Space Shuttle Columbia

One of the most recent ones and also regarded as one of the most tragic mishaps in the history of space travels is that of Space Shuttle Columbia. Just like Space Shuttle Challenger, Columbia also had onboard a seven crew members which included Captain Rick Douglas Husband, Pilot William 'Willie' Cameron McCool, Payload Commander Michael P. Anderson, Payload Specialist Ilan Ramon, and Mission Specialists David McDowell Brown, Kalpana 'KC' Chawla, and Laurel Blair Salton Clark. After lifting off from the Kennedy Space Center on January 16, 2003, Space Shuttle Columbia successfully completed its 16 day mission in space carrying out various experiments on microgravity and loads of other experiments in SPACEHAB research module. The mission STS-107 was the 28th mission of the spaceshuttle Columbia. While NASA prepared for the return of Columbia to Earth after accomplishing a successful mission on 01 February 2003, none of those present there at the Kennedy Space Center would have expected anything unusual to happen during the landing. But the shuttle was only 16 minutes away from the Kennedy Space Center, when the unexpected event took place and the spaceshuttle Columbia was suddenly seen disintegrating into several pieces. It is said that over 85,000 pieces of debris from the spaceshuttle fell from the skies over Texas. The nation and the world lost all the seven crew members onboard Columbia. The reason for the failure was later found to be a hole on the left wing that allowed atmospheric gases to bleed into the shuttle as it entered the Earth, leading to the loss of the sensors and the spaceshuttle Columbia itself. Once again NASA was forced to suspend its spaceshuttle flights for over two years.

□

5
The Road Ahead

Exploring the unknown, pushing beyond the limits, the quest to know and control all that is not even visible is what that makes the mankind the ruling species on the earth. Scientists are hopeful that their recent achievements in space operations, especially the success of optimistic Mars missions will soon make human race a two planet species.

Rediscovering the Moon

NASA's new vision of the future space program include, taking manned missions to our nearest celestial neighbour – moon by 2020, which will be precursor to the proposed visits to Mars and beyond. Russian space agencies too are working on the same lines. However, this time winning the space race is not the motive behind the efforts. The collaborative effort of the human race as such and not of individual countries is what will pave the road ahead to space.

Why space programs include revisiting the moon, a cosmic body which astronauts have already visited?

NASA strategists offer an explanation to this; according to them the experience of moon missions would be helpful in planning exploratory programs of more distant worlds. Lunar stays of longer duration will help the astronauts and scientists build upon the experience gained while planning other long-term space missions that include visiting farther planets. Revisiting the moon and extended stays on its surface will also help astronauts discover the possibilities of using moon as the base of operations where humans can replenish essential supplies including oxygen, rocket fuel that may be created using the resources available on moon. If humans have to reach deeper space then development of such skills is absolutely essential.

The mission named the Constellation Program also takes into consideration accomplishing certain near-term scientific objectives. These scientific mysteries include exploration of the investigation of water ice near the moon's poles. Before sending manned missions to the moon space scientists plan to send robotic spacecrafts to the lunar surface in search of the landing sites which can offer the best of the available resources to astronauts. Scientists believe that the south pole of the moon is particularly promising for the human stay because it is rich in hydrogen and gives an indication of the presence of water ice.

NASA has designed a new state-of-the-art spacecraft, the Orion capsule, which the NASA officials describe as "Apollo on steroids," owing to its similarities to the Apollo spacecrafts but with much bigger size and modernity's introduced in the design. The new capsule is larger, at least three times in volume capacity and the more space on the inside enough for a four person crew. NASA engineers have developed a safer and reliable design in the Orion capsule. The new technology makes the Orion different and safer than the familiar shuttle now used by NASA. Upon reaching the space the flexibility incorporated into the design would allow the same Orion vehicles to ferry astronauts to and from the International Space Station. Astronauts would be able to enter the lunar orbit from where the landers can repeatedly visit the moon's surface. Just like shuttles the Orion capsule may be reused up to ten times at least. The capsule will land using parachutes, just like the earlier spacecrafts, however the new design allows them to touch the dryland safely rather than take a splashdown in the ocean. Engineers and NASA experts believe that after the success of Orion capsules they can assemble Mars-bound vehicles in orbit to take the first humans to the red planet.

Terraforming: Making Other Solar Planets and Far Off Galaxies Habitable for Human Race

NASA engineer Adam Steltzner said, "Exploration and the curiosity that motivate it are fundamentally human." Steltzner was addressing the audience at "The Future is Here Festival", organized by *Smithsonian* Magazine in Washington, D.C. The festival held in May 2014, gave a platform to various radical ideas about how humans one day would not only discover but also live in the alien worlds that are today studied from a distance. Today these ideas may appear more like science fiction than facts but the optimistic human race is nothing without some dare and a few dreams. Some private industries such as 'Mars One', have already started working on a project that seeks to establish a permanent settlement on the Red Planet. Steltzner previously worked in the capacity of the lead engineer for NASA's Mars rover, 'Curiosity'. He helped to design and test the rover's one-of-a-kind descent system.

A faction of scientists believes that for humans to reach other planets and spread even beyond the solar system, landing on various celestial bodies is not the only way. Speaking at "The Future is Here Festival", Steltzner also suggested that exploration of space can be carried out in multiple innovative methods.

He said, "Imagine hurtling durable terraforming bacteria to another world, conceivably with the idea of shaping that body...."

The idea that bacteria – a primitive life form present on the earth – could always hitch ride aboard various manned and unmanned space missions to other planets is not new. Steltzner suggested a well-designed program to implement this idea. According to him these bacteria could carry out our genome and the instructions to reassemble it after landing on a planet. The assumption made here is that the planet is already terraformed to support life as we know it.

Terraforming may be described as the hypothetical process of deliberately modifying its atmosphere, temperature, surface topography or ecology to be similar to the environment of Earth to make it habitable by Earth-like life.

According to Steltzner this is like "Printing human beings organically over time."

Acceptance of such programs on the conven[illegible] parameters of success in space exploration is yet debatable. Steltzner pointed out that humans are much more than their genes, and that our experiences and connections play an important role in shaping us. Nevertheless, in terms of spreading humanity through the galaxy, such seeding might be the easiest and most effective.

Terraforming projects may appear to be a long distant reality, visiting planets in the solar system remains a not-to-distant possibility. "Technology didn't slow us downgetting to the moon; technology won't slow us downgetting to Mars," said Steltzner adding to his views on the proposed theory.

The 100 Year Starship

National Aeronautics and Space Administration (NASA) and U.S. Defence Advanced Research Projects Agency (DARPA) have jointly commissioned a project "The 100 Year Starship (100YSS)", in October 2014. The project aims at developing tools and technology required in building and flying a spaceship that can make interstellar travels. However, the main focus of the 100YSS is to create a business plan spanning over 100 years that would help in fostering the research needed for space travels. The two major agencies of the USA funded this project to involve future generations in the research and development of the breakthrough technologies that will eventually help in achieving the goal of interstellar travels. The long-term project is not looking at just travelling to our near neighbours in solar systems but to other galaxies too.

According to Mae Jamison, the American physician and former NASA astronaut, who heads this project, believes that since the mission is possible to be completed in a [illegible], it is important that a system is created that future generations carry forward developments and achievements made by the previous associates of the project. The program isn't necessarily concerned with building the ship itself as much as it seeks to foster innovation and enthusiasm for interstellar travel.

According to Jemison, "The reason we're not on the moon has nothing to do with technology and everything to do with public will and commitment..." "It has to be an inclusive journey."

"The 100 Year Starship", aims at increasing public enthusiasm for space programs. The program includes engineers and astrophysicists, artists, science fiction writers and all those who agree to synergize their efforts in the exploration of the deep space.

The project faces criticism from many people. The U.S. Senator Tom Coburn called, "The 100 Year Starship", as one of the 100 most wasteful government spending projects. Critics believe that funding the space programs, such as the 100YSS is extravagant when there are humanitarian needs that have to be met on Earth. Jemison's counter explanation to this allegation is that such exploration often leads to innovation and unexpected technology that make an impact on Earth-based programs.

In words of Jemison, "I believe that pursuing an extraordinary tomorrow will create a better world today...."

To get there, Jemison emphasized that everyone must be involved in the process.

Common people are not any more as enthusiastic about the space exploration and discoveries as they were when astronauts first forayed into space. The sole reason for this may be attributed to the communication between the specialists and the layperson.

"The public did not leave space," Jemison said while discussing the reduced enthusiasm. "The public was left out of space."

Mars and Beyond: Space Travels of the Future

Louis Dill Friedman, an American astronautics engineer and space spokesperson in his recent book–Human Spaceflight From Mars to the Stars, discusses about the distant future of human space travel. Friedman is not only a dreamers but also a qualified aerospace engineer, who knows exactly how dreams may come true especially when it comes to making various manned and unmanned missions to space a success.

He elaborates upon two ideas for successful space travels and explorations.

The first one is called, the Asteroid Redirect Mission. In this mission astronauts and scientists would take hold of a large piece of asteroid and move to another location, which is convenient enough for astronauts to reach. According to Friedman NASA is already working on the project and will achieve a breakthrough within a decade; by relocating an asteroid to a reachable location and making it a literal stepping stone on the way to the journey to Mars.

The second idea revolves around sending a robotic mission into deep space towards the focalline created due to the bent in sun's light; the bent in which case is caused by a planet out of the solar system. Theoretically speaking solar gravity lens focus is capable of magnifying planets by at least billion times. Thus building a 10-centimeter telescope in there would be equivalent to an 80-kilometer telescope on the earth. The robotic missions would thus be able to get high-resolution images of the potential land in deep space which can support human life as we know it. According to Friedman, at least one can think of such missions being successful in our lifetime.

Discussing the above two ideas, Louis Friedman said in an interview with the magazine *Discover*, "These two missions illustrate the point of my book nicely: Human space flight nearby, robotic spacecraft far away, extending the human presence together."

In the same interview he added:

"Human exploration and colonization of Mars will keep us busy for hundreds, even thousands, of years. During that time, there will be advances in nanotechnology, space sailing, robotics, biomolecular engineering, and artificial intelligence. These advances are occurring even now, affecting our outlook about what it means to be human and engage in human activity. Those technologies will not merely allow us to stay home on Earth and Mars, but our minds will extend our presence throughout the universe so that we will not need or want to extend our bodies there – even if we could, which I think is doubtful."

Scientists, space engineers, astronauts, fiction writers and even the common men put forward several theories about probability of successful space missions. Some of these may be a distant dream and some achievable but time consuming projects. Yet others and a reality. Even though the ethical dilemma of spending insane amount of money on dream projects instead of improving the life on the planet confronts the decision-makers. Space enthusiast believe that the road ahead would lead the human race to a distant world which is full of promises.

□□□